Rivers by Design

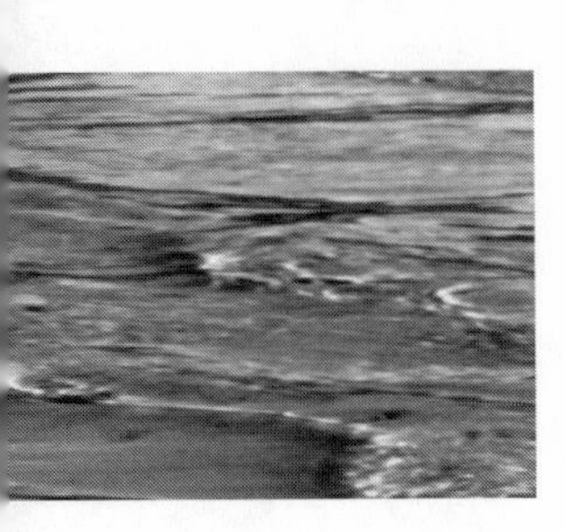

DUKE UNIVERSITY PRESS *Durham & London* 2006

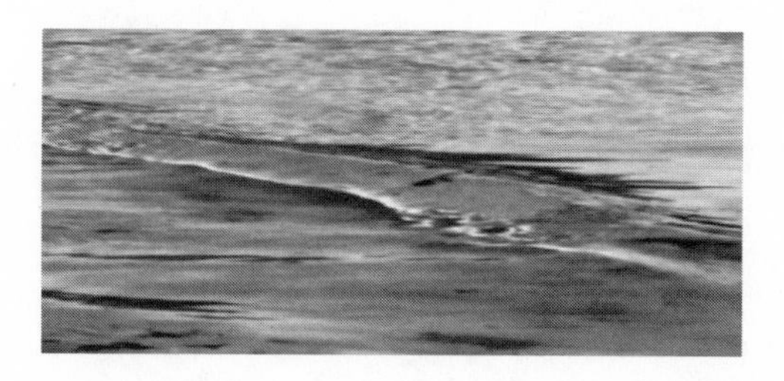

KAREN M. O'NEILL

Rivers by Design

STATE POWER AND THE ORIGINS OF U.S. FLOOD CONTROL

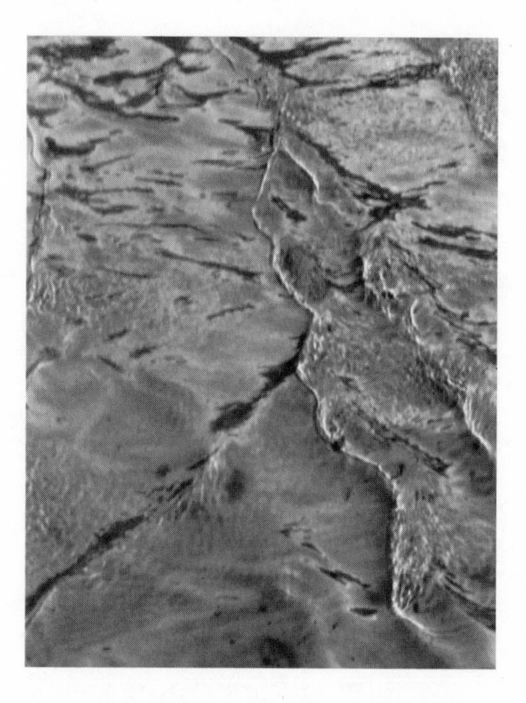

Printed in the United States of America on acid-free paper ∞
Designed by C. H. Westmoreland
Typeset in Scala with Arepo display by Keystone Typesetting, Inc.
Library of Congress Cataloging-in-Publication Data appear
on the last printed page of this book.

for Alex and Evan

Contents

Tables and Maps

Preface

What is a river? Standing on the bank of a river in most areas of the United States, we are likely to think about rainfall and runoff, the landscape's slope, reeds, fish, and insects as the sorts of features that define a river. Accustomed to a domesticated landscape, we have trained our eyes to look past seemingly small modifications like sandbags or plantings that shore up banks along wooded river reaches. At the same time we may have stopped thinking of concrete-lined channels below highway bridges as rivers at all.

Engineering projects and legal agreements have remade nearly every river in the United States and have shaped expectations around the world about how governments should control the environment. Standing next to a high dam, for instance, we are impressed with the engineers' ability to channel and store a river and may even begin to think of that river primarily as a source of electricity. Low-profile works for flood control are not typically included in sightseeing tours but have rebuilt many more of the United States's rivers. The Los Angeles River was made into a concrete flood channel that carries rain water to the ocean with such speed that it is usually dry enough to be used for Hollywood car chases. The upper Mississippi River is a series of still pools bounded by navigation locks and flood control levees, earthen berms built alongside rivers to confine the flow. And the historic "river of grass" that flowed south from Lake Okeechobee to the Everglades has been diked and channeled, allowing farmers to drain wetlands and plant sugar cane.

These and other rivers simplified by flood control structures defy our commonsense understandings about rivers but have been tolerated as environmental sacrifice zones. Flood control structures wash out river eddies where fish feed, block the flow to wetlands, and reduce the amount of fresh water available for local human use. As we rethink our decisions to control rivers by debating whether to "naturalize" river channels and restore wetland flooding, we are also considering how to restructure our relations to government.

This book analyzes the social origins of the United States's flood control program, which represents one possible institutional solution for managing rivers. I particularly aim to explain why the program was designed to link the central (federal) government with local and subnational state government institutions, including landowner-run levee districts. This involves asking who pressed for this program and why the pattern of articulation emerged.

With the United Nations estimating that two to seven billion people will lack ready access to fresh water by 2050, there is growing interest in understanding the institutions that manage our water. About 60 percent of the world's largest rivers have been altered by hydraulic structures—including flood control works—that limit our flexibility in planning for future ecosystem and human needs.[1] Because water projects in the United States have inspired many of these structures, studying the social elements of water engineering in the United States may yield lessons about the prospects for achieving economic and political development goals elsewhere. The social origins and consequences of the United States's irrigation, hydropower, and drinking water supply projects are increasingly well documented. Programs for land drainage, wetland filling, and river flood control are less well understood, even though they have likewise yielded wealth and political power for some.

The United States's flood control program is primarily directed by the federal government, but it was initiated by elites from two outlying regions, not by power-seeking government officials. Planters, shippers, and merchants from the lower Mississippi Valley who were seeking to make their region more competitive with the established northeast originally defined aid for flood control as a program for economic development. They gained indirect assistance from Congress beginning in the 1820s, in the form of river surveys and swampland grants. Farmers, merchants, and investors in northern California began demanding flood control aid for the Sacramento River system in the late 1850s, after hydraulic gold mining operations in the Sierra Nevada range intensified the flooding of valley farms.

Farmers and city builders in the riverfront areas of these two river valleys had already built some levee lines privately and as members of local levee districts, because they could not drain their lands without first blocking these rivers' characteristic floods. Advocates for federal flood control aid used these levee districts as their political base for organizing. They formed alliances with downstream merchants, passed subnational state government levee regulations, and promoted the issue of flood control within their home regions. They also campaigned for federal aid by organizing river conventions and gaining support from their members of Congress. By 1900, they had built regional and national lobby organizations that worked with business leaders, civic groups, and elected officials from river areas throughout the country to lobby Congress for flood control aid to the Mississippi and Sacramento valleys.

Congress unofficially directed the Corps of Engineers to repair and improve Mississippi River levees in the 1880s, at a time when the central government was still quite small. Most voters and politicians at that time felt that the U.S. Constitution restricted the federal government to aiding the interstate distribution of goods, not their production. Northern voters were especially unlikely to support aid benefiting planters in formerly rebellious southern states. Until 1917, this work was therefore justified publicly as benefiting river navigation, rather than as flood protection for riverfront lands. After years of lobbying by flood control activists, Congress created an official flood control program for both the Mississippi and Sacramento rivers in 1917 and then extended the program to all navigable rivers in 1936. In designing this program, members of Congress and the Corps of Engineers expected the Corps' existing system of field offices to oversee construction. These officers in turn hired local contractors and worked with subnational state and local government flood control agencies to plan specific works. This pattern of implementation articulated the three levels of government and involved landowner beneficiaries in program decision making. It represents one pattern of administration that has become characteristic of the United States. As historian Martin Reuss remarks, given the cultural and political tensions between central and local power in the early republic and the creation of "a republican form of governments within the government. . . . It is little wonder that [the United States] saw no successful implementation of co-ordinated public works administration. Perhaps more surprising is that this became a permanent condition in the United States."[2] This lack of coordination was evident in the inadequate and confused response to the devastation of Hurricane Katrina in 2005 in southern Louisiana.

As a study of the social origins of the flood control program, this book is neither a history of legislative power politics nor a history of the Army Corps of Engineers' flood control program. My goal is not to explain why or when officials approved specific flood control bills or projects, decisions that depended on vote trading in Congress, on social or environmental crises, and on broad programs of state building like the New Deal. Scholars have written legislative histories of the major flood control bills and of bids to create comprehensive plans for multipurpose water use, such as the Tennessee Valley Authority (TVA). I cite these in the narrative.[3] I also rely on scholars who have detailed some of the local political developments that led to flood control activism.[4] Major elements of the story of social origins, however, have not been documented in the research literature. This book presents new data from information produced by the river activists themselves, describing the themes they used and their methods of organizing. It provides an overview of flood control advocacy through 1936 to consider how this advocacy produced a program that linked the implementing agency, the Corps of Engineers, with landowners and with subnational state and local agencies.

Efforts to alter the course of rivers have changed social and political structures in expected and unexpected ways. The lower Mississippi and Sacramento valleys received the first federal flood control aid. Along the lower Mississippi, planters were able to expand cropping onto formerly malarial swamplands. Although planters had been unwilling to risk the lives of slaves to work those lands, they encouraged freed African Americans and recent European immigrants onto those lands after federal levees were built. These areas remained susceptible to levee breaks, but sharecropping and other forms of labor control passed the risk of flooding onto these laborers. The flood control program encouraged the commitment of planters and merchants to the goal of reviving the cotton economy and achieving economic autonomy from the North, a goal that eluded them. In the Sacramento Valley, federal court intervention and the promise of federal flood control aid ended a political stalemate between gold miners in the Sierras and valley farmers, whose lands had been inundated by debris from hydraulic gold mining in the mountains. These federal government actions effectively ended the Gold Rush, but they also fostered the rise of intensive specialty agriculture in the Sacramento Valley portion of California's Great Central Valley. In other river valleys throughout the country, flood control levees, reservoirs, and spillways have similarly helped farmers to devote their resources to intensifying production and have walled off cities from their rivers.

These river projects committed the government to developing specific rivers for specific purposes, altered the risks of living alongside rivers, and deepened the government's responsibility for managing uncertainty and responding to disasters. The earliest forms of federal assistance for flood control were discussed as forms of aid that would improve economic security by protecting farming areas and ports. Over time, people built residences and businesses in the floodplains behind river levees. Government levee building became increasingly focused on protecting population centers.

But the rivers push back. As more water is impounded behind upstream levee lines and shifted through diversion channels, and as more people build in downstream floodplains, the damage that is likely to result from a break in the levee system is greater than ever. Property losses from floods have increased dramatically since the government began building flood control works because levees and floodwalls made floodplains seem safe for building. The federal government now has floodplain management programs and coastal protection programs to improve ecological protections against floods. It has failed to take a more direct approach by requiring that local governments prohibit development in the most vulnerable floodplains once federal levees and floodwalls have been built. The federal system of government therefore complicates the task of responding to the dynamic interaction of weather, land, and rivers in landscapes that we have altered.

The country's political and cultural fragmentation also makes river projects politically vulnerable. Advocates of government reform have long complained that the lists of river projects approved by Congress are fueled by pork barrel politics. Environmental critics add that these projects have been environmentally devastating as well as wasteful. The unprecedented damages of the 2005 hurricane season provoked a new round of debates about the social, financial, and environmental costs of federal projects for river and coasts.

As the following chapters show, river development politics led to changes in relations between the regions and the federal government. As the original national highways, rivers were first manipulated by local and subnational state governments. The federal government slowly took on its own river projects. In part I, chapter 1 presents scholarship about relations between regional elites and the modern state, considering how a centralizing state organizes outlying territories. Readers interested in the story of the flood control campaign, rather than in theory, may wish to skip to the second chapter.

Chapter 2 explains why decisions about land development and infrastructure set some of the federal government's basic domestic powers during the republic's early years. Federalists, National Republicans, and Whigs—most notably John Calhoun and Henry Clay—argued that internal improvements would create common interests by physically uniting the country and facilitating commerce. Democratic critics painted these plans as examples of governmental overreach. When the Supreme Court ruled in 1824 that the Constitution's interstate commerce clause made the federal government responsible for free access to river transportation, Congress directed the Army Corps of Engineers to begin a program to improve river navigation channels. This court case became the key constitutional justification for central government intervention into the economy, although for decades funding was usually limited to aiding the interstate distribution of goods, not their production. Because flood control work would directly enhance agricultural production by allowing farmers to control soil moisture, it was not acceptable under this view of the Constitution.

The rise of flood control activism in the Mississippi and Sacramento valleys is discussed in part II. Chapters 3 through 7 detail why leaders in the Sacramento and Mississippi river valleys were the first to seek flood control aid in response to early federal commitments to infrastructure building.

Because the massive Mississippi River gives access to the heart of the continent, settlers from the United States pressed for the federal government to take the Louisiana territory from the French, to improve the river for shipping, and to protect riverfront lands from floods. From the time the French settled New Orleans in 1717, local authorities had organized levee districts run by landowners to coordinate local levee building along the lower Mississippi River. Under United States rule, large landowners, politicians, bankers, and shippers from port cities all along the Mississippi organized river conventions from the 1840s on to protest the disproportionate share of federal river and railroad aid going to the northeast. Many river activists from the lower Mississippi also began to argue that the river's volume and meandering ways caused flooding as well as navigation problems and that the two problems should be solved together. As a low-cost gesture typical of the times, Congress approved grants of federal government swamplands to a dozen subnational state governments west of the Appalachians beginning in 1849. The subnational states were to sell these lands to subsidize flood control works. Even so, southerners' continued resentment over the uneven allocation of rail and river navigation aid contributed to sectional polarization.

Territory might have been permanently lost to the United States after the Civil War, but instead, the central state was remade. As Barrington Moore argues, once the central state was no longer required to protect slavery to maintain the Union, it could devote itself more fully to economic development policies, with the support of a southern elite newly accepting of government economic interventions.[5] In their post–Civil War memorials to Congress, pamphlets, and editorials, southern river activists expressed the belief that federal promotion of northern industry could be accompanied by renewed promotion of the southern cotton economy. Many of these writings implied that the government owed aid to the South as war reparation. Recipients of the 1840s swampland grants (whom Populists derided as swamplanders) joined with merchants and other boosters in the 1870s to organize new river conventions demanding federal flood control aid. Members of Congress from the lower Mississippi Valley gained leadership of House and Senate river committees and traded votes over the years to win navigation projects. Finally, Congress informally directed the Army Corps of Engineers in the 1880s to use navigation program appropriations to assist levee districts on the lower Mississippi.

The integration of California into the union was less fraught, but it too provoked conflicts about what the government owed to new territories. Within five years of the discovery of gold along the Sacramento

River in 1848, miners shifted to intensive methods such as hydraulic mining, which used water under pressure to blast away hillsides in the Sierras. Tens of thousands of yards of rocks and clay flowed into the river system and soon onto the lands of valley farmers, who supplied the miners. As farmers and townspeople gained some economic independence by selling their goods outside of the mining areas, they began to protest mining damage.

Federal officials and judges were presented the choice of intervening between these two vital industries in California, a step unprecedented at the federal level. Even by the late 1860s, after editorialists and judges had widely acknowledged the damages from mining, Californians remained divided over the best response. Miners and many mining-dependent valley residents admitted that the harms from mining were visited upon downstream property owners but they asserted a principle often expressed in case law that defended mining as a higher and more productive use of land. Recipients of large swampland grants, like their counterparts in the Mississippi Valley, favored demanding that the federal and state governments build flood control works. Residents of valley areas dominated by small farms tended to prefer lawsuits and injunctions against mine owners, suspecting that miners would use government flood control works to justify more mining. In the end, lawsuits worked more quickly than politics did. In what was likely the first major environmental ruling of a federal court, a judge in 1884 extended case law protections against harms by ordering hydraulic mine operators to build and maintain vast debris basins. This requirement pushed the costs of mining beyond the reach of most operators. The lengthy ruling also documented the federal government's failure to regulate mining claims and to prevent harms.

The former adversaries regrouped. They now argued that the court ruling obliged the federal government to aid miners and farmers. Aid would include building debris dams and river improvements throughout the river system, in conjunction with existing California state flood control measures and with local levee districts. Congress created a commission to investigate mining damages and to plan improvements.

Courts and Congress had generally limited the federal government to facilitating the interstate distribution of commercial goods. Interventions into California's mining dispute and levee aid for planters on the lower Mississippi River were instances of federal aid to economic production that encouraged further interventions in the twentieth century.

Part III analyzes the role of river activists in fostering the rise of a system that distributes most federal development aid through regional competition. Chapters 8 through 10 discuss the unification of the two regional campaigns into a national campaign seeking aid for the Mississippi and Sacramento valleys and describe their success in passing formal flood control bills. Flood control advocates from the two valleys had begun forming a joint campaign at the turn of the century fronted by professional lobby organizations, which attracted support from commercial organizations throughout the country. The Mississippi River Levee Association lobbied for aid to the lower Mississippi. Advocates also eventually convinced the National Rivers and Harbors Congress, the influential river navigation lobby, to add to its wish list flood control aid for the Mississippi and Sacramento rivers. The 1917 Flood Control Act directed the Corps of Engineers to provide levee aid for the Mississippi and Sacramento rivers. In 1928, after immense and deadly floods on the Mississippi, Congress committed the federal government to wholly redesigning the Mississippi and Sacramento river systems to control floods.

Chapter 10 analyzes the final step ensuring that government water development in the United States would be fragmented across special purpose programs, like the flood control program, rather than coordinated through comprehensive planning. Scholars of political development argue that the reformist New Deal broke the nineteenth-century pattern of limited government. New Deal dollars certainly expanded land and resource programs, but the flood control program and most other resource programs retained their orientation. President Franklin Roosevelt and other reform-minded New Dealers pushed to create nonpolitical, professionally managed organizations such as the TVA to centrally plan and supervise multipurpose river development projects. The New Dealers' eagerness to act did provide opportunities for advocates of reform in resource policy, and they managed to pass the TVA bill. But it also provided opportunities for advocates of the reform-resistant flood control program. Despite the early promise of the TVA, Roosevelt's signing of the national Flood Control Act of 1936—which empowered the Corps of Engineers to build projects for all navigable rivers—

effectively ended nationwide, comprehensive planning for natural resources at the federal level. The 1936 Flood Control Act institutionalized the influence of river interest groups and the articulated pattern of implementing federal projects in cooperation with local institutions. It also marked the triumph of the "river lobby" in ensuring that the Corps of Engineers would retain management over nearly all navigable rivers through its navigation and flood control programs. Other water management duties would be divided among agencies including the TVA, Bureau of Reclamation, Soil Conservation Service, and Bonneville Power Administration. In the process, these government programs have created new physical vulnerabilities for people who settle in areas that appear to be safe.

Chapter 11 looks back at the development of river policy to consider how earlier ideas and policies about rivers helped shape governance and changed the ways we use rivers. The pattern of articulation between the three levels of the federal system and between the government and landowners allowed for a workable division of labor for designing, funding, and building flood control works. The goals of these interrelated institutions are not easily changed. The Corps of Engineers' navigation and flood control programs and their state and local partners have resisted financial reforms and have addressed the ecological problems caused by engineers by proposing ever more ambitious engineering works to restore waterways. Infrastructure works and the human organizations that created those works continue to impose limits on our actions and our imagination.

Acknowledgments

For comments on this research over the years, thanks to Richard Berk, Lee Clarke, Margaret FitzSimmons, Bill Freudenburg, Jess Gilbert, Bob Gramling, Gregory M. Hooks, Ivan Light, Michael Mann, Rachel Parker-Gwin, Benita Roth, Dolores Trevizo, Andrew P. Vayda, and Maurice Zeitlin. Norris Hundley Jr., Donald Pisani, Alex Ridgway, William G. Roy, and Thomas Rudel read the entire manuscript at various stages and provided me lasting advice about thinking and writing. Anonymous reviewers for the journal *Rural Sociology* and for Duke University Press gave exceptionally useful suggestions for framing my argument. Camille Bevere, Stella Capoccia, Adam Diamond, Buck Foster, and Jamie MacLennan helped with administrative and research assistance. At Duke University Press, thanks to Sharon Parks Torian, Mark Mastromarino, former editor Raphael Allen, and especially to J. Reynolds Smith for his calm assurances and his insights about why we write and read books. Michael Siegel became a partner in this project by creating maps showing that conflicts over river engineering at times pitted upstream against downstream and North against South. Diane Bates, Bob Boikess, Caron Chess, Kerry Ferris, Mary Gibson, Bill Hallman, Colleen Hatfield, Rachel Holmes, Alisa Lewin, Rebecca Li, Rick Ludescher, Bonnie McCay, Marla Pérez, Fiona Stevens, Dan Van Abs, and B. J. Walker provided professional support. I am grateful to my grandparents who farmed in California's San Joaquin Valley, Germano and Claire Borba, for inspiring a lifelong interest in the ways we use land and to my other family members, who have never asked why the book took so long to finish and who supported me in so many ways. Research for this book was partially supported by a UCLA fellowship and by the New Jersey Agricultural Experiment Station.

A portion of chapter 10 appeared in different form as "Why the TVA Remains Unique" in *Rural Sociology*, 2002.

part I

Rivers and State Authority

Infrastructure building during the early nineteenth century contributed to several related processes of social change that increased the central state's power and defined how the state would relate to citizens. The United States government organized its territory, established a unified property system, and aided the rise of the national market and the industrial sector. Chapter 1 presents questions about territory, federalism, and regional activism that guide this study. Chapter 2 outlines how those processes unfolded during the early years of the United States as decisions were made to develop rivers. These chapters lay the groundwork for parts II and III, which describe how activists and their legislative allies from the Mississippi and Sacramento valleys got Congress to increase development aid by increments.

chapter 1

Infrastructure Builds the State

The flood control system built by the United States Army Corps of Engineers is rivaled only by the system protecting the Netherlands. But while the Netherlands could scarcely exist without river levees and seawalls, the United States has plenty of land outside of floodplains.[1] It is not obvious why the United States government would take on the responsibility of providing flood control and flood insurance for lands along all major rivers.

As later chapters will explain, demands for federal government flood control aid by landowners, shippers, financiers, and politicians from the Sacramento and Mississippi river valleys made the local and regional problem of flooding into a national responsibility. Far from being imposed by the central government, the flood control program was resisted by Congress and the Army Corps of Engineers. Activists first established laws and public works programs at the local and subnational state government levels to assist private flood control work. Once the federal program was created, federal managers had to work closely with local contractors, levee districts, and subnational state governments to build federal levees and weirs. This activity articulates the central government with local and subnational state government institutions, and it articulates the government with landowners. Organizational articulation is one possible institutional form that affirms central government authority in regions that are physically distant from the capital, even though in the case of flood control such institutions were not imposed from the center.

This chapter outlines the broad processes of modern state building and class formation affecting the way the U.S. government manages land and resources, setting a conceptual framework for analyzing why the pattern of articulation emerged in the flood control program. Establishing and sustaining territorial power is a defining feature of modern states. Studying relations between the central government and distant regions is one way of considering how the physical integration of territory contributes to modern state power.[2] Centralizing authorities typically repress internal challengers and set up border garrisons and administrative controls to manage outlying regions. Effective rule depends, however, on a wider range of activities, including economic development projects, changes in the law, and discursive work. These activities produce institutional forms that manage interactions between the central government and outlying regions. In the case of flood control, they produced an institutional form that articulated the federal, state, and local governments.

People in outlying regions influence the nature and timing of activities that build government power, and they even initiate efforts that end up affirming central government control. Like many other land and resource programs in the United States, the flood control program was created because provincial elites demanded aid. Landowners, shippers, and merchants from the lower Mississippi and Sacramento river valleys argued that the federal government had a duty to control floods that threatened valley farming and shipping and that hindered participation in the national economy.

Two features relevant to the control of outlying territories are special to this case, namely that it concerns land policy and that it unfolds in a country with a federal system. Land, resource, and infrastructure policies often produce visible symbols of central government power that become essential for sustaining daily life in a locality. These policies also require modifying legal and social systems that regulate access to land. Studying a federal system highlights how the authority and power of a central government can be extended by responding effectively to local demands and by incorporating local institutions.

These features of the flood control case draw our attention to the organizational, legal, and cultural boundaries between the modern state and society, rather than to the central government's bureaucracies, budgets, or armies. The flood control program is what George Steinmetz calls a "structure-changing policy," one which alters the way subsequent policies are produced by altering the perceived boundaries between the modern state and society.[3] The flood control program altered boundaries by redefining the government's political duty to assist landowners, while giving the impression that it was merely ensuring some morally prior landowner right to property that is ready for productive use. In particular, politicians and judges in the early nineteenth century had interpreted the interstate commerce clause of the Constitution as limiting federal river work to navigation projects that facilitated the distribution of goods. By contrast, flood control projects would directly improve the way goods were produced (especially crops), not just the way they were distributed, and would directly benefit landowners.[4]

Flood control activists and sympathetic officials did not set out to change the nature of property and the state. Neither did they anticipate that the path to success would involve temporarily defining levee repair as navigation work, trading votes for regionally specific development aid in Congress, calling for national rather than merely regional flood control aid, emphasizing public safety rather than economic development, and making delicate political tradeoffs with Progressives and New Dealers.

Activists did consciously build on the long-standing assumption in the country's culture that the government should support rather than threaten the institution of property. The Mississippi carried much of the nation's water, they pointed out. The Sacramento was burdened by debris from mining that had built the nation's gold reserves, which were considered at that time to represent the country's wealth. Flooding was therefore a national problem that unfairly burdened landowners. The nature and timing of specific political steps leading to federal aid depended on large-scale political and economic conditions. The cultural and political transformation of local and regional flooding into a national problem, however, depended on the links between federal, state, and local governments and between government agencies and landowners established over the decades while they worked to change rivers. To explain why this sort of transformation helps to define the modern state, the next sections consider how territorial power is sustained domestically, how modern state territorial claims and property laws regulate access to land, how these institutions affect the formation of landed classes, and how federalism structures space.

The Modern State and Territory

Modern states organize territory by reorienting social networks toward state activities.[5] In the United States this involved changes in property rights and changes in the central government's legal and physical organization of territory.

Scholars readily agree that modern states differ from other forms of political power because they organize territory more intensively. In a statement cited by theorists of many persuasions, Max Weber defines the modern state as "a human community that (successfully) claims the *monopoly of the legitimate use of physical force* within a given territory."[6] Before the full emergence of modern states in Europe, religious, military, and political authorities typically had jurisdiction over specific groups of people, rather than authority over well-defined territories.[7] Historically, the emergence of a territorial power in a region forced other powers to organize to protect territory, or face possible annexation.[8] By gaining authority to draw resources from a specific territory, a state may become able to finance territorial expansion or to create more intensive political controls within its existing territory.[9] Political expectations, nationalist ideologies, international agreements and norms, international aid packages, and targeted military actions have fostered the organization of the world into territorial states.[10]

Some observers argue that international governmental organizations, transnational nongovernmental organizations and corporations, and border-spanning economic activities and technologies have eroded states' roles as authoritative political and economic actors, or at least changed the conditions under which modern states operate.[11] Writings on globalization invite us to consider whether modern states have until now been as unified and territorially well defined as is often assumed.

Scholars who view state power as being imposed from the center have contributed to the idea that the modern state is unitary. Tocqueville argued that the French state not only centralized power and resources but also compelled the provinces to adopt a national culture.[12] Many studies inspired by Tocqueville[13] describe regional resistance and acknowledge that modern state power is often indistinguishable from the power of local landed elites, but the focus of these studies remains on central bureaucracies.[14]

Others question the assumptions that modern states are monolithic organizations distinct from civil society that impose order and political culture from the center.[15] Critics of Weber's definition of the state note that states often fail to monopolize coercion but remain recognizable as states in their attempts to control territory.[16] Empirical studies find that the actual practice of state sovereignty varies, to the point where challengers within the borders of some countries have established themselves as alternative regional authorities.[17] And nationalities seldom fully coincide with national borders, even when state leaders encourage nationalist movements.[18]

Peter Sahlins reconsiders assumptions about state control over outlying territories by studying a border area of a country strongly associated with centralized control, France.[19] Ethnic Catalans on both sides of the emerging border with Spain resisted the two centralizing states, while using their new national identities and nationalist claims to territory to compete against each other locally for scarce resources. Sahlins concludes that political links between French Catalans and the French government were built therefore from the provinces as well as from the center, despite the Catalans' long-standing ambivalence toward Paris. Sahlins's approach provides a model for investigating initiatives from outlying regions in the United States that determined how the central government related to people in these regions as it managed rivers.

Stein Rokkan identifies two fundamental domestic conflicts relevant to such struggles for territorial control that mark the rise of the modern state: (1) a conflict between centralizing state regimes and resistant peripheral regions and (2) an industrial revolution that stimulates a class cleavage between owners and workers and a sectoral cleavage between landowners and industrialists.[20] Rokkan's two conflicts unfold in part as struggles over the way land is organized and used. Using these concepts, a historical case study of state building would analyze events to consider how specific institutions were established to manage each of these two forms of conflict.

As Timothy Mitchell sees it, state building is not accomplished by leaders imposing their will on the people to suppress these two conflicts. Instead, it occurs through social processes that change not only how people organize control over material conditions but also how they perceive the institutions that control those conditions. The state appears to be a separate entity that regulates society. Despite this appearance, case studies of policy making show that the border between the state and society is ambiguous and ever changing. For instance, governments may privatize or nationalize industries, take on old age care, and create incentives for industries to regulate their own activities. Rather than conceiving of the state as an entity that imposes order over territory, Mitchell therefore proposes that we investigate how mundane practices of spatial organization, such as border guards and watchtowers, lead us to perceive of an authoritative modern state set apart from society.[21] Sahlins and others who study borderlands have used this approach to investigate the social and cultural processes that (provisionally) transform frontiers into international borders.[22] To understand how land is regulated *within* the borders of an advanced capitalist country, I consider how Mitchell's dividing line is drawn between state territory and private property, the two key institutions regulating land in such countries.

The Modern State and Property

In addition to excluding other sovereigns from their territory, modern states also regulate access to land by their own citizens. With the rise of capitalism, states have managed a mix of capitalist, usufruct, and other forms of land claims. In the United States, people have relied on the language of law when debating decisions about "internal improvements," that is, government-sponsored land development projects and public works. Within the Anglo-American tradition of law, according to Edward Levi, decisions about internal improvements and other types of land development are treated as answers to "the perennial problems of government: the relationship between problems of the person, the state, and property rights."[23]

Putting this in theoretical terms, state territory is a form of authority that overlaps with private property on most land in the United States. Anglo-American law represents this overlap by conceptualizing property as a bundle of rights that is divided between landowners and the state. The fee simple estate in the United States includes the exclusive right of the landowner to hold land and its permanent features, the rights to use it or dispose of it, and the rights of freedom from interference or damage by others.[24] Society, represented by the state, always withholds the right to tax private property, the right to condemn land for public use (eminent domain), police power, and the right to reclaim land if the owner fails to maintain ownership rights (escheat). Government may also regulate land use indirectly with its spending power.[25] The modern state's coercive role is built into these rights, and the actual benefits that owners enjoy vary greatly as circumstances change and as political and judicial decisions are made. Property rights are often evoked politically in the United States as a prior right that *checks* state power.

At the same time, the state presents itself as the legal *originator* of property rights. Property in the United States actually incorporates a jumble of historical claims, including squatters' claims, parcels designated for indigenous peoples, and land grants from colonial authorities, the federal government, and subnational state governments.[26] By acting to guarantee property rights, no matter what their origin, the state not only mediates between conflicting private parties but also actively defines the terms of these social relationships.[27]

Property rights establish landowners as a group that depends heavily on government decisions. The state's reach into claimed regions in turn depends on state relations with landed interest groups, in their roles as key players in the land tenure system (i.e., landowners, under capitalism), and as elites who depend on the development of their home regions.

The Modern State and Formation of the Capitalist Landed Class

Because modern states coexist with a variety of class and market systems, it is clear that there is more than one path of class formation and market development that enables states to establish some measure of territorial control.[28] Countries with capitalist economies, however, do experience broadly similar changes in the social uses of land that affect both state building and class formation. Geographers conceptualize the transformation of land into state territory and private property as the transformation of highly differentiated "places," defined by use values, into planned "space," where exchange value and state plans predominate.[29] When the government enforces contract and property law or builds roads, it transforms the distinctive qualities of places into plots usable for market or state uses.

Emerging groups of agricultural capitalists might target several factors of production and distribution as they organize to influence law and land policies, including policies that convert places into space.

Capitalist agricultural production depends on having the institutional means for owning and exchanging property, a ready supply of workers, the legal and social means for extracting labor, physical and legal access to markets, and control over growing conditions.[30] Max Pfeffer identifies three systems of farming in the United States that satisfy these factors differently.[31] After the Civil War, farming in the Sacramento and lower Mississippi river valleys—where flood control activism originated—evolved into two of Pfeffer's three systems, corporate farming based on wage labor (in California) and sharecropping and tenancy (along the lower Mississippi). Despite the great differences between these two farm systems, landowners in these two valleys—which feature large, flood-prone rivers—organized in similar ways to demand flood control aid.

This suggests that the pressure to increase production under capitalism inclined well-off farmers to seek governmental environmental controls, no matter how else they satisfied the other factors of production. The federal flood control program eventually came to feature levees and emergency channels and basins that divert high-water flows and transform selected floodplains into economically useful space. These measures increase the market value of many parcels along a river, redistribute risk within a river system (e.g., by placing the highest levees around cities), and protect ports that link a river to the national market.

Agricultural capitalists and their eventual allies did not necessarily anticipate or desire all of these outcomes, but they did demand environmental control measures that seemed likely to aid capitalist class formation and to ensure their regions' influence in national politics. The resulting government programs enhanced patterns of uneven development across the country and so prompted new political demands for aid. In the United States, competition between the regions plays out through the federal system.

Federalism and Territory

It is difficult to determine where sovereignty lies within federal systems, and so regional competition often centers on defining which level of government can exercise specific powers. Regional leaders constantly make and change bets about whether centralizing certain powers will benefit their home region. Historically, writings about federalism in the United States have emphasized interpretations of the Constitution, talking alternatively about the federal government's coordination of subnational states or about the subnational states' independence from the federal government.[32] Nineteenth-century political players felt constrained by the Constitution, but many found gains by asserting their own creative interpretations. As constitutional debates became less salient in politics, scholars came to define federalism through the political actions that shift authority among the three branches and among the national, subnational state, and local government levels.[33]

Political practice as well as theory therefore recognizes that the state in the United States includes the three levels of government, not just the central government. Localism is inherent in this system, where the Constitution assigns residual powers to the subnational states and where nearly all elected officials run for office from geographically defined districts. Because members of Congress have their home districts in mind when voting on implementation plans, and because local institutions have well-established legal and social powers, growth in central government powers may also increase local and subnational state government powers, rather than simply shift power to the center. This dynamic makes it especially attractive for regional politicians to seek federal aid.

Conclusion

The rest of the book explains the rise of the flood control campaign in the context of central government efforts to integrate territory under its control and investigates how the articulated set of flood control institutions emerged in the federal system. Part I concludes with chapter 2, which describes specific military, legal, and political conditions in the newly established country that would affect the flood control campaign. Working within the federal system, flood control activists would later seek to expand the role of the army in developing the trans-Appalachian West and to broaden interpretations of the 1824 Supreme Court decision that held the federal government responsible for free access to rivers. They also argued that a policy benefiting a specific region is in the national interest because people of that region share rights and duties with others in the nation.[34] These early years established that debates about the dividing line between private property and the state would be critical in defining the domestic powers of the state.

Parts II and III describe the organization of agricultural production into two distinct farming systems in the Sacramento and lower Mississippi river valleys and the rise of flood control campaigns in each valley. Elites in these two valleys demanded flood control aid in response to Stein Rokkan's two types of conflicts, conflicts between regional elites and centralizers and conflicts between rural landowners and rising industrialists. Regarding the conflict over centralizing, single-minded and persistent organizing and politicking for flood control aid helped elites in the Sacramento and lower Mississippi valleys to direct federal resources to their home regions and to gain some sense of control over relations with the central government.

Rokkan's second type of conflict manifested in the Sacramento and lower Mississippi valleys as a struggle to make local agriculture more competitive in an industrializing economy. In both valleys, those who depended on agricultural trade competed for government aid against the manufacturing-based northeast, and they competed for market share against agricultural producers in other United States regions and abroad.[35] In the view of activists, flood control aid would improve river navigation and enable farmers to intensify production on fertile lands by helping farmers regulate field moisture throughout a growing season. Flood control also helped planters along the lower Mississippi River to secure a postslavery labor force by opening new lands to sharecropping. Farmers in California succeeded in creating a new, industrialized form of agriculture, while farmers in the lower Mississippi River area did not.[36]

Like the Catalans,[37] flood control activists in the United States used their national identity to solve local problems. Activists argued that aid to their two rivers was in the national interest. They attracted allies from other regions on that basis and on the basis that flood control aid could become a tool for developing these other regions. These efforts began in the provinces, not in the capital, but they led to creation of a new central government program that articulated its powers with local and subnational state institutions. As this case suggests, each of Rokkan's two conflicts unfolds as a struggle over the way the modern state's claimed space is culturally, politically, and physically organized.

chapter 2

The Founding Principles of River Development

Essayons [Let us try]—Army Corps of Engineers motto

Early conflicts over rivers led to decisions treating rivers as public resources managed by the federal government, rather than as commodities or as resources that should benefit some states over others. These decisions defined and redefined federal government relations to the states and to landowners, setting legal and ideological principles that would guide relations to territories the country acquired in later years.

It was at meetings called in 1786 to resolve violent interstate disputes over oyster beds and shipping tolls that James Madison and Alexander Hamilton proposed what was to become the final constitutional convention. Their intention was to grant the new United States government sufficient power to mediate these and other interstate disputes.[1] Demands by settlers from the United States for access to the French colonial river port in New Orleans set into motion negotiations that led to the Louisiana Purchase in 1803. And in 1824, a Supreme Court ruling in *Gibbons v. Ogden* declared it the duty of the federal government to prevent states from limiting interstate commerce. Congress immediately authorized the Army Corps of Engineers to improve specific navigation conditions on rivers. The army engineers' navigation program soon became the most active program of central government intervention into the economy.[2]

In the context of these events, boosters in the distant territories west of the original thirteen states began to promote federal government river development as a means for promoting regional growth, while insisting that most control over resources should still reside at the local level. Conceptions of the border between the state and society at that time also allowed for the central government to regulate the interstate distribution of goods but did not accept federal aid for the production of goods.

Who Claims the West?

As a settler colony, the United States formed its borders, property system, class system, national markets, and governance structures while industrial capitalism was emerging. Even before they founded the United States, business, legal, and political leaders in the colonies were conscious that their decisions about internal improvements would shape these emerging institutions.[3]

The compromises that changed the United States from a confederation into a stronger union thrust the new "general government" into decisions about land and resource management. Seven of the original thirteen states made overlapping colonial claims to lands from the Appalachian range west to the Mississippi River. Leaders of the six other, smaller states were therefore reluctant to federate as weak partners.[4] Following difficult debates from 1783 to 1787, the seven states agreed to surrender to the central government their most dubious western claims.[5] Congress was to have these public lands surveyed, ensure property rights, and distribute plots to private owners to encourage orderly settlement and passage to statehood.[6]

Acquiring and incorporating these lands contributed to long-lasting regional conflicts. Some easterners considered it unfair to grant western public lands directly to the new western states, rather than selling the lands to benefit all states.[7] Many southerners believed the admission of new, nonslave western states would ruin the balance in Congress between slave and nonslave states.[8]

Through decades of political party debates and petitions by would-be land recipients, Congress directed federal agencies and states to distribute most of the public lands to private owners.[9] Some remaining lands were organized as Indian lands or as national parks or other conservation areas. Other lands were granted for capital cities, schools, jails, canals, or other public works.[10] These policies, together with decisions about rivers, established private rights and the public good as two poles that would define legal and political discussions about land and resources.

A National Economy Begins to Emerge

Decisions defining property rights[11] and decisions about distributing public lands heated up two broad questions: Would the central government actively develop the economy? Would central government actions favor northern industries or southern export-oriented slave plantations?[12] Politicians were keenly aware of inequalities between the dominant northeast and the other "sections."

The sections came to depend on each other soon after the western territories were settled, and federal government actions were critical in creating this emerging national economy. Federal government land grants for settlement and infrastructure building served as capital in the new western states and created a taxable base of landowners.[13] But as early as the 1790s, settlers in Kentucky, Tennessee, and the Ohio Valley began to argue that land grants were not enough. To avoid paying bribes at the Spanish seaports of New Orleans and Mobile, settlers had developed supply routes via northern rivers and lakes to import crafts and export furs and agricultural goods. The most vocal of these settlers complained of their need to structure their own supply routes and asserted that the United States should have to earn their loyalty, not assume it. They demanded that the government acquire New Orleans so that they might ship their goods directly down the Mississippi.[14]

Imbalances between western areas and the industrializing northeast grew as the southern economy grew. In 1793, Eli Whitney invented a cotton gin capable of processing the short-staple, upland cotton grown in the southeast. Textile manufacturing technology and labor processes were adapted from England, and Francis Lowell applied river power in his northeastern facilities for cotton processing and textile manufacturing. Coupled with increased trade and defense spending for the War of 1812, these changes transformed the states' economies. The United States government—and New York merchants who gained control over the cotton trade to England and France—began to exercise new power in trade relations. Plantation slavery was invigorated. By 1800, cotton production had pushed the country's economy beyond agrarian self-sufficiency, enabling northeastern bankers and merchants to finance ventures in infrastructure building and manufacturing in the northeast.[15] An awareness of sectional inequalities conditioned every vote in

Congress and every presidential decision about infrastructure legislation. This awareness was nurtured in the ideological debates that created the first political parties.

Internal Improvements and the Parties

Ideologists cultivated different philosophies about federal assistance for "internal improvements" to distinguish their parties from each other. Party pronouncements in the 1820s left tensions between the sections implied. By the 1850s, these were replaced by statements actively cultivating sectional division.

Federalist Alexander Hamilton set an ambitious plan to create currency controls, to shift some of the tax burden onto farms and away from businesses, and to boost infrastructure. Hamilton and other Federalists supported selling most of the public lands to replenish the Treasury, and they favored creating a lighthouse board and making river and harbor improvements to provide services vital to the general welfare. According to Hamilton, improvements to river navigation are authorized by the Constitution's commerce clause, which provides Congress with the power to "regulate Commerce with foreign Nations, and among the several States, and with the Indian Tribes."[16] During the 1830s, the American Whig party supported similar interpretations.

Federalist plans provoked Thomas Jefferson and James Madison into formally organizing the Anti-Federalists into the opposition Republican faction in 1794, the predecessor to today's Democratic party. Jeffersonian Republicans included small farmers throughout the country as well as members of the early Tammany machine in New York City.[17] The Jeffersonian Republican party favored a "strict construction" of the Constitution that would give the central government limited fiscal powers. They took this stance with the intention of sustaining the liberal credit terms that farmers enjoyed under the laws of several of the states. Jeffersonian Republicans preferred granting land in small parcels to farmers and spending revenues on only a small number of essential, interstate public works.[18]

Both Federalists and Anti-Federalists envisioned the central state as an institution that could address the failings of other institutions like the national market or the national culture. Of particular importance to river development was the argument of Federalist party leaders that vigorous central state programs of internal improvements linking the sections would create common interests across the sections. Jeffersonian Republicans, by contrast, held that town and countryside had fundamentally conflicting interests. The state should instead work to create common interests by creating a nation of yeoman farmers. Disagreements about tactics and about the definition of political unity continued to color political debates long after these early parties disbanded.[19]

Despite differing ideologies, most party leaders, when in office, directed tangible benefits to their home regions. In practice, Federalists and Jeffersonian Republicans alike approved a number of development measures in addition to land grants, although Republicans did restrict project spending.

Aid for Development Begins

As president (1801–9), Jefferson had a mixed attitude about internal improvements. He opposed a postal road that would have benefited New England Federalists. He approved giving western states a small portion from public land sales to finance public works such as the Cumberland Road. Many such projects were finished only after Congress and the president quietly granted cash appropriations from the federal Treasury, an action quite contrary to Jefferson's Republican ideology. He opposed a military elite, but he revived the Army Corps of Engineers, which Congress had first formed as a unit in 1779. He also signed bills to have the army build piers, harbors, and lighthouses for civil purposes.[20] By approving land grants for school building, Jefferson opened an ideological loophole for strict constructionists, who thereafter considered direct money grants unconstitutional but land grants acceptable.[21] Jefferson's Secretary of the Treasury, Albert Gallatin, pushed further by proposing in 1808 a grand program of surveys, canals, and frontier highways to support farming. This was popular in the West, but some politicians thought it would require a constitutional amendment, and Congress defeated it.[22] And as discussed in chapter 3, Jefferson's most dramatic act to promote growth was his approving the extraconstitutional purchase of the vast Louisiana territory from the French in 1803.

Proposals for an American system by Henry Clay and John Calhoun galvanized debate in Congress for fifteen years.[23] After pushing President James Madison (Democratic-Republican, 1809–17) into the War of 1812, expansionist war hawks Clay and Calhoun floated plans to promote western development featuring tariffs, a central bank, and internal improvements. These bills drew solid support from the western delegation but vetoes by Madison.[24]

Politicians continued to shift between approving politically pragmatic projects and debating the principles behind government interventions. These debates led to the next party system. Representatives from the Mississippi Valley won federal land grants to support canal building in some areas.[25] During James Monroe's administration (Democratic-Republican, 1817–25), Congress approved expeditions to the West, river surveys, improvements on lighthouses and piers, and assistance from the Corps of Engineers in building the Cumberland Road.[26] In the 1820s, the Democratic-Republicans split. Jeffersonians who were impressed that the War of 1812 had stimulated the industrial economy joined with former Federalists to become National Republicans, predecessor to today's Republican party. One of their intentions was to increase government intervention in the economy.[27]

Roads, Canals, Railroads, and Federalism

Before 1820, goods were moved within the United States on small, muddy roads built by farmers under the guidance of local officials, on privately built turnpikes, on a handful of short canals, through coastal waters, or by river.[28] By the early 1820s, state governments were building so many transportation projects that federal government officials felt little need to take the lead.[29] Optimism abounded about the potential for native capitalists to finance works with the aid of state governments.

The 363-mile Erie Canal was fantastically difficult to build, but toll revenues repaid bond holders within seven years.[30] Overcome with enthusiasm, state and local officials throughout the country chartered dozens of mixed enterprises to build canals, granted them loans, provided them technical advice, bought company stock, and granted them powers of eminent domain and rights to collect tolls.[31] Between 1830 and 1860, Congress also granted over 4 million acres of public land for canals, and it spent $3 million for canal company stock.[32]

Competition to build connections to the West and to intensify networks in the East soon involved government-assisted railroads. To compete with the Erie Canal, Baltimore merchants and city officials raised public and private funds and got the Army Corps of Engineers to survey a route for the country's first passenger railway in 1831, the Baltimore and Ohio Railroad.[33] When it became evident that towns and ports might wither if bypassed by a rail line, states and local governments mounted great efforts to raise money for new rail lines.[34] Some of the canals and most of the railroads succeeded. Others were engineering failures, went bankrupt, or were underused or abandoned.[35]

Transportation links—including steamboats that began serving the lower Mississippi River in 1816—were sufficiently extensive to promote a "market revolution" between 1815 and the 1850s. Farmers in all regions shifted acreage from crops grown for barter to crops grown for sale on the emerging national market.[36] As transportation improved, policies encouraging international trade and industrialization in the northeast made it physically feasible and economically sensible to shift the major grain growing areas from the northeast to the upper Mississippi Valley. Easier access to international ports and to produce from the Midwest encouraged southern planters and farmers to shift from food crops to cotton. As a result, the market revolution deepened the social differences between the slave south and the free-labor north.[37]

By the 1840s, the industrialized northeast had better transportation links than any other region did. Southeastern states and New England states had spent less than half the amount that mid-Atlantic states had spent on canals and roads in the 1830s. Ventures in mid-Atlantic states also attracted most of the private investments in infrastructure.[38] After the Panic of 1837 ruined many mixed-enterprise ventures, calls for federal government aid increased.[39] Federal aid was not politically inevitable. The legal justification for providing aid, however, was already in place.

Judicial Power, Federalism, and Rivers

The canal boom at its height inspired a 1824 Supreme Court decision that defined the waterways as public, set conditions on federalism and on government relations to landowners, and cleared the way for the central government to build river and harbors projects.[40] Early courts had adapted the ready traditions of English law to set many of the conditions for market exchange and for governance. In *McCulloch v. Maryland* in 1819, the Supreme Court established that the "necessary and proper" clause of the Constitution justified the priority of central government supremacy over the states. In *Gibbons v. Ogden* in 1824, the court made clear that this power could be substantial.[41]

The court held in *Gibbons* that the Constitution prohibited New York State from granting steamboat monopoly rights on the state's rivers. The monopoly was held by steamboat inventor Robert Fulton and his backer, Robert Livingston, who shared another monopoly on the lower Mississippi River. Representing the steamboat operator who filed suit against the monopoly, Daniel Webster argued successfully that states cannot interfere with Congress's constitutional authority to regulate interstate commerce.[42]

Gibbons defined only a potential. Illegal toll taking and piracy took years to police, and states and private companies continued to manipulate canal rates, and, later, railroad rates. But this first important interpretation of the commerce clause established a constitutional foundation for a national market, a goal favored by Chief Justice John Marshall. It also made the interstate commerce clause the main constitutional tool for justifying federal government interventions into the economy, allowing the federal government to directly regulate private property rights.[43]

"Rivers and Harbors" Pork Barrel

Members of Congress were so impressed by the Erie Canal that they had already been debating two river bills. The *Gibbons* ruling settled lingering doubts about the constitutionality of improving river navigation. One uncontroversial bill, which became the General Survey Act, began a permanent program of river surveys and authorized the army to lend engineers and surveyors to private canal and railway companies. The Survey Act attracted support from the public and from legislators by including projects for many localities, becoming the model for legislative logrolling.[44]

Eastern legislators fought the second bill, which listed river improvement projects. Western legislators had passed an earlier bill directing the army to conduct the first comprehensive survey of the Ohio and Mississippi rivers. In their 1822 report, army engineers confirmed travelers' tales that the Ohio was only three feet deep and that the lower Mississippi held thousands of submerged logs. The engineers recommended that the government devise new methods to remove snags, to build dikes and dams that would narrow the river and erode sand bars, and to build levees—earthen mounds alongside the river—to hasten navigation and to keep the river from flooding.[45] Legislators and voters commonly regarded it unconstitutional for the government to build works benefiting private landowners. So legislators included in their bill all projects but the levees.

A representative from Kentucky argued that the Ohio and Mississippi rivers were the common stock of the country and the federal government's responsibility, because they ran between, not within, the western states. The House passed the bill. Several East Coast senators organized opposition on constitutional grounds, but the bill passed with a slight majority.[46] This 1824 bill became the first Rivers and Harbors Act, appropriating $40,000 for two pier projects and $75,000 for experiments on sand bars on the Ohio and for snagging on the lower Mississippi River by the Army Corps of Engineers.[47]

Why the Corps of Engineers?

The Army Corps of Engineers built both civil and military works. From the revolutionary period on, the army had fortified eastern harbors and coasts, in part to protect civilian ships.[48] The withdrawal of European rivals then provided Congress an exceptional opportunity. For decades after the 1812 war, Congress devoted the army to exploring the West, to waging wars of attrition against indigenous peoples, and to surveying and building public works.[49] Military leaders liked to point out that civilian shippers also benefited from military transportation works in the West. For instance, before the steamboat period, keel boat crews who returned by foot from New Orleans to Pittsburgh traveled on the army's Natchez Trace.[50] By 1824, Army outposts were established throughout the West, and the army had the government's only body of skilled builders. The river program would become the army's largest civilian project.

Those who mistrusted a standing army or opposed planning by experts argued against the river program. Other politicians cited ideological reasons for supporting army infrastructure projects. According to Hamilton and the Federalists, the Constitution's provision for a peacetime military implied that Congress may use the army to direct public works. Jefferson suggested that civil works would keep soldiers employed, yet at readiness for military action, and would encourage discharged soldiers to undertake peaceful pursuits.[51] The promise of aid for the home district was appealing for pragmatic political purposes as well. Over the years, the Corps would build thousands of projects that satisfied constituents.[52]

The Corps became the first federal agency to actively manage land and resource development decisions involving the central government, local business people, and local officials. During work on the Philadelphia harbor in 1828, congressional public works committees, army engineer officers, and local business and shipping interests began to work cooperatively and to resist outside interference. This pattern would characterize the waterway program for over a century. In the 1840s, some Corps snagging and dredging crews, faced with erratic congressional funding, raised funds from local interests to continue their work. In Chicago, New York, Baltimore, Milwaukee, and other cities, local commercial groups and city officials organized volunteer crews or contributed funds and supplies to sustain army waterway projects. One army engineer financed a dredger by selling excavated sand and by leasing out the dredger. Army engineers also sold equipment, moved funds from office accounts into project accounts, and solicited testimonials to Congress from shippers and commercial interests detailing deteriorating waterway conditions.[53] In dealing with the problems of implementing projects, then, the Corps unofficially engaged in local, state, and national politics.

Internal Improvements under Jacksonian Democracy

During the late 1820s and 1830s, Congress reduced funding for military engineers, passed appropriations irregularly, and distributed the pork as widely as possible. Legislators approved relatively inexpensive, but temporary, dredging operations more readily than expensive structural projects such as canals or dams. Over one hundred federal water projects remained unfinished in 1836, and projects slated for the West failed to satisfy western boosters.[54]

Members of Congress were unenthusiastic about proposals by National Republican John Quincy Adams (1825–29) to create a new patent system, a university, an observatory, a department of science, a naval academy, and an improved army engineering academy. Each state did receive aid for rivers, harbors, or canals. Army engineers also followed the advice of Adams and Henry Clay to publicly promote the cause of internal improvements.[55]

President Andrew Jackson (Democrat, 1829–37) rejected Adams's nationalistic ideology in some of his statements but signed onto many projects in answer to petitioners from the West.[56] While Jacksonian Republican Democrats often questioned internal improvements, they were most concerned with preventing elites from taking control over government development.[57] Congress passed appropriations for waterways nearly every year between 1824 and 1838. A Georgia senator called the 1836 act a "monster." In disdain of army professionalism, Jackson and the Democrats in Congress directed funds to local rivermen, such as steamboat engineer Henry Shreve, especially to party loyalists. Although Jackson denounced the "corrupting influences" of public works, he signed over $3 million in project appropriations, notably including work on a portion of the Cumberland River leading up to his own plantation.[58] Presidents Martin Van Buren (Democrat), John Tyler (an errant Whig), and James Polk (Democrat) likewise publicly criticized internal improvements while signing a favored few waterway projects.[59]

Cost overruns on waterways projects, abandonment of the federal government's troubled Chesapeake and Ohio Canal excavations, civilian engineers' complaints that the army threatened their livelihoods, and project failures at the Philadelphia and Erie harbors and at the mouth of the Mississippi River prompted hearings in the House in 1836. The Chief of Engineers was indicted in 1838 on findings that War Department accounts had been mismanaged.[60]

This ended the period of active spending on rivers begun in 1824. Congress banned private employment of army engineers, budgeted only $2,000 for waterways in 1840, and split $485,000 among twenty-three water projects in 1844. Corps officials themselves dropped certain controversial projects. They also created the basic tools of cost-benefit analysis to determine whether a project would be of local or national significance, assessing project costs, project feasibility, and the locality's economic potential.[61] With states and private firms bearing overwhelming debts for infrastructure projects after the Panic of 1837, and with cuts in federal spending, little capital became available to complete projects until the late 1840s.[62]

Nation-Unifying Public Works Return, Briefly

Democratic President James Polk (1845–49) set imperial adventures against internal improvements by vetoing a small river appropriation in 1847 during the Mexican War. But members of Congress were so impressed by wartime engineering feats that they assigned the army engineers to make surveys for a transcontinental railway, to build new defensive works, to improve navigation on the California rivers, and to survey flooding and navigation problems on the Mississippi River.[63] Protests against Polk's 1847 veto took different forms in northern and southern states, as discussed in chapter 4. This marked the end of the assumption that advocates of river projects were all seeking the same ends.

In the fluid party and sectional politics of their era, the most nationalistic members of the Whig Party (created in 1834) had little chance of creating the national government that they envisioned, which would guide morals and boost the economy. The Whigs did, however, crystallize sentiments for internal improvements, especially among those who stood to benefit from economic development guided by the central government.[64] The Democrats approved platforms for presidential elections from 1840 to 1852 asserting that the federal government had no constitutional power to finance and build a general system of internal improvements. But Democrats in Congress approved lesser interventions.[65]

The parties shattered and reorganized in the 1850s over the question of preserving slavery, and internal improvements became less salient in partisan politics. During the early years of the United States, parties had been socially heterogeneous organizations that built alliances in Congress across the sections. During Jackson's administration, politicians had begun to reorganize parties to pursue sectional interests. By the 1850s, parties were largely split by section.[66] As a result, party members became more intent on gaining internal improvements for their own sections than on using internal improvements to distinguish their parties ideologically. The South became almost solidly Democratic by 1852, and in the 1860s the Democrats split into northern and southern organizations. The Whig party, which had attracted most southern Louisiana sugar planters, faded after most of its southern members left. Northerners formed the new Republican party in 1854 to oppose slavery and to preserve the Union.[67]

In the 1850s, both the Democratic and the Republican platforms favored land grants for a transcontinental railway, and the Republicans favored tariffs and appropriations to build railways and waterway projects.[68] Congress made the first land grants for railroads, and the Senate created a joint committee with the administration to plan internal improvements.[69] The 1852 platforms for the Whig Party and the antislavery Free Democratic Party asserted that the Constitution vested in Congress the power to improve navigable rivers and harbors to provide for defense and for interstate commerce projects. Then, after the 1850s, internal improvements disappeared from national party platforms.[70]

As discussed in chapter 4, northern Republicans' approval of internal improvements for northern states in the 1850s only enhanced the sectional split. As northeast transportation improved, southerners increasingly interpreted poor conditions on the Mississippi River and inadequate southern rail links as indicators that their region had lost autonomy under the United States government, especially as northeastern cities tightened control over the cotton trade.

Conclusion: Central Government Intervention and Landowner Relations

As decisions were made about land and rivers, the federal, state, and local governments gained specific and lasting powers that would set limitations on later policy decisions. In the context of the evolving federal system, landowners, merchants, and officials began to organize their local economies and their congressional delegations to compete against other locales for position in the national economy. Although the pork barrel would become a steady feature of nineteenth-century congressional politics, members of Congress did not merely distribute spoils. They also shaped property rights and exercised federal government powers identified by the courts. Members of Congress and judges would spend the rest of the nineteenth century interpreting the 1824 *Gibbons* ruling in particular and the rightful powers it gave the federal government to regulate and aid economic activities, including the development of private lands. The prevailing interpretation of that ruling before mid-century held that it would allow for navigation channel and port improvements in the interest of preserving the distribution of commerce between the states but that it would not support improvements to land such as flood control works that directly improved the production of commercial goods. The border between state and society appeared to place production entirely in the hands of landowners or industrial producers, although the occasional hushed cash appropriation from Congress suggested the possibility of greater government assistance for production.

According to public land historian Paul W. Gates, Congress's decision in 1787 to retain federal authority over the public lands within western states gave the federal government means for exercising its sovereignty and stimulated the growth of government and civil institutions within the new states.[71] By contrast, the original thirteen states retained their own authority over public lands when the United States was founded. Citizens and officials in the western public land states have since that time been motivated to mobilize various coalitions to demand land grants as well as financial and technical aid to develop that land. As will become clear in later chapters, decisions shaped by conditions in the West became generalized to the entire country.

part II

Regional Competition and the Rise of the Flood Control Campaign

Because of the strategic value of rivers, decisions about rivers helped determine how the trans-Appalachian territories would be incorporated into the United States. The campaign for flood control aid began in two places, the lower Mississippi and Sacramento river valleys. Boosters argued that the U.S. government had a responsibility to facilitate commerce and settlement along western waterways if it was to claim those territories. They succeeded in establishing a campaign that drew support from outside their two regions. By the end of the nineteenth century, leaders of the campaign could assert with growing conviction that aid to their two regions would be in the interest of the entire country.

New Orleans, Chicago, and San Francisco had become centers of development in the new territories because they were the meeting places of inland and international waterways. They retained that status during the nineteenth century through concerted economic and political efforts to build infrastructure and attract investors and settlers. Military access to the interior and economic access to trapping grounds, fertile lands, and mineral deposits made these port cities strategically important. Rival colonizers competed for sovereignty in the trans-Appalachian West. They built state powers, settled supporters around these city sites, and engaged in military exploits and diplomacy.

Acquiring these territories changed the federal government's relations to foreign governments and to the subnational states. New questions about the supremacy of the central government also arose. Southerners, midwesterners, and westerners demanded development assistance for the new territories. Northeastern elites demanded forms of aid that would reinforce northeastern dominance. The land and re-

TABLE 1. Origins of Flood Control Activism: Similarities and Differences

	Mississippi River	Sacramento River
Strategic location	—Sovereignty: U.S. challenges indigenous peoples, France, Spain, England —Interstate river —Upstream agriculture and industry, downstream international port city	—Sovereignty: U.S. challenges indigenous peoples, Spain, Mexico —Intra-state river —Upstream mining and agriculture, downstream international port city
Rural land tenure	—Large plantations since colonial period —African slave labor, then tenant labor	—Large holdings since colonial period —Low-wage immigrant labor and tenancy
Central state policies toward region	—Navigation improvement aid —Large-holding pattern reinforced by land policies —Civil War	—Navigation improvement aid —Large-holding pattern reinforced by land polices —Weak presence during Gold Rush
Regional political development	—Landowner-run levee districts empowered by states. Districts spread under swamp land grant program —Flood control advocates join merchants and officials advocating navigation —New Orleans–dominated lobby demands aid for the Mississippi, then for the nation	—Landowner-run levee districts empowered by states. Districts spread under swamp land grant program —Uneasy alliance of large mine owners and farmers, organized by merchants —Activists demand federal aid for the Sacramento, then join national lobby

source programs that Congress eventually created to administer federal policies were small at first, but they established a pattern of close cooperation between resource agencies and local program beneficiaries.

Historians have often treated relations between the federal government and the South separately from relations between the federal government and the Far West. The southern Mississippi Valley had slavery, was sparsely settled, and had colonial elites who persisted after the

Louisiana Purchase. The Far West adopted free labor and attracted settlers who formed cities, and its colonial elites were pushed out of landownership. Even so, there were many similarities in the way the federal government managed land in the two regions. Federal government policies produced important similarities in the land tenure systems and land politics of the South and the West as they developed capitalist agricultural bases.

Part II shows that state making and class formation processes in the lower Mississippi and Sacramento river valleys gave rise to two remarkably similar elite pro-growth movements. The lower Mississippi and Sacramento valleys produced two of the three major systems of farming that Max Pfeffer identifies in the United States, sharecropping along the Mississippi and corporate farming based on wage labor along the Sacramento.[1] Although these two systems were produced through quite different regional social and political processes, there were many similarities in the ways they interacted with the federal government. Flooding was a dramatic feature of both valleys. Capitalist farmers in both valleys allied with downstream merchants and shippers in port cities to demand similar environmental controls. These were the first two valleys to create large flood control activist movements and the first two valleys to receive federal government flood control aid (see table 1, above).

Congress made land grants that created large landholdings in these valleys. It also sent Corps crews to improve navigation for the military and for civilians. In both valleys, largeholders, local officials, and business persons who depended on local production and shipping called conventions to demand more federal aid for rivers.

The following chapters use activists' documents to describe the rise of flood control activism and the origins of working relations between the Corps, members of Congress on river committees, and local elites. These relationships shaped development in the two valleys and defined a pattern of state growth that became characteristic of other federal government land programs.

chapter 3

The Mississippi River

BECOMING THE NATION'S RIVER

> One who knows the Mississippi will promptly aver—not aloud but to himself—that ten thousand River Commissions, with the mines of the world at their back, cannot tame that lawless stream, cannot curb it or confine it. . . . But a discreet man will not put these things into spoken words; for the West Point engineers have not their superiors anywhere.—River pilot, Mark Twain, *Life on the Mississippi*

The Mississippi as a Key to Western Expansion

The most important strategic objective as the United States expanded west was gaining and sustaining access to the lower Mississippi River. This objective stimulated central state growth and set land and resource development as a long-term priority in domestic politics. The Louisiana Purchase, the Civil War, and the reconciliation of North and South after that war pushed the U.S. government to integrate, reclaim, and reintegrate southern lands. These three events also became entangled in decisions about river use and river development. Local pro-growth advocates repeatedly demanded federal development aid to boost the region's economy and enable it to integrate into the national market. This chapter describes the establishment of U.S. rule in the lower Mississippi Valley, settlement, the rise of the slavery-dependent cotton economy, and relations between landed interests and the central government that laid the groundwork for the flood control campaign. Chapters 4 and 5 continue the narrative, describing the rise of the flood control activist movement, the contribution of river politics to Civil War tensions, and the first instance of federal aid for flood control in the 1880s.

Campaigns for regional development, such as the flood control campaign, arose in response to U.S. expansion to the West. Even before any of the trans-Appalachian territories were organized as states, settlers there demanded central government political and economic aid. As this chapter explains, during the early nineteenth century, leaders of the emerging slavery-based cotton economy on the lower Mississippi began to concentrate their political demands on making the river a highway that would allow New Orleans to control the United States's cotton trade with Europe.

Although planters and members of Congress from the lower Mississippi river area were wary of powers that could enable the federal government to regulate slavery, they did not necessarily reject economic assistance. Many planters sustained strong Whiggish sensibilities in favor of federal government intervention, at least until the Democratic and Republican parties polarized by sections in the 1850s. This point is important in that it indicates that ideas and practices defining the boundary between the state and landowners have a complex history and that it has not been a matter of moving from minimal to increased government intervention. As chapter 2 indicates, few in the North or South saw government as existing wholly apart from enterprise during the early years of the Republic. One measure of planters' sensibilities before the Civil War is their reliance on local and state government levee programs, as explained below. These institutions coordinated private efforts with the work of levee districts, which socialized the costs of flood control and built a local political base for promoting the flood control enterprise.

Mississippi, the Nation's Waterway: The Physical Setting

Features that made the Mississippi River susceptible to flooding also made it attractive for settlement. Government policies encouraging settlement were in turn used by activists to insist that the government therefore had an obligation to control floods. The Mississippi River is the second largest river in the world, draining 41 percent of the continental United States and providing access to great portions of the interior.[1] Historically, the river deposited tons of topsoil on its floodplains,

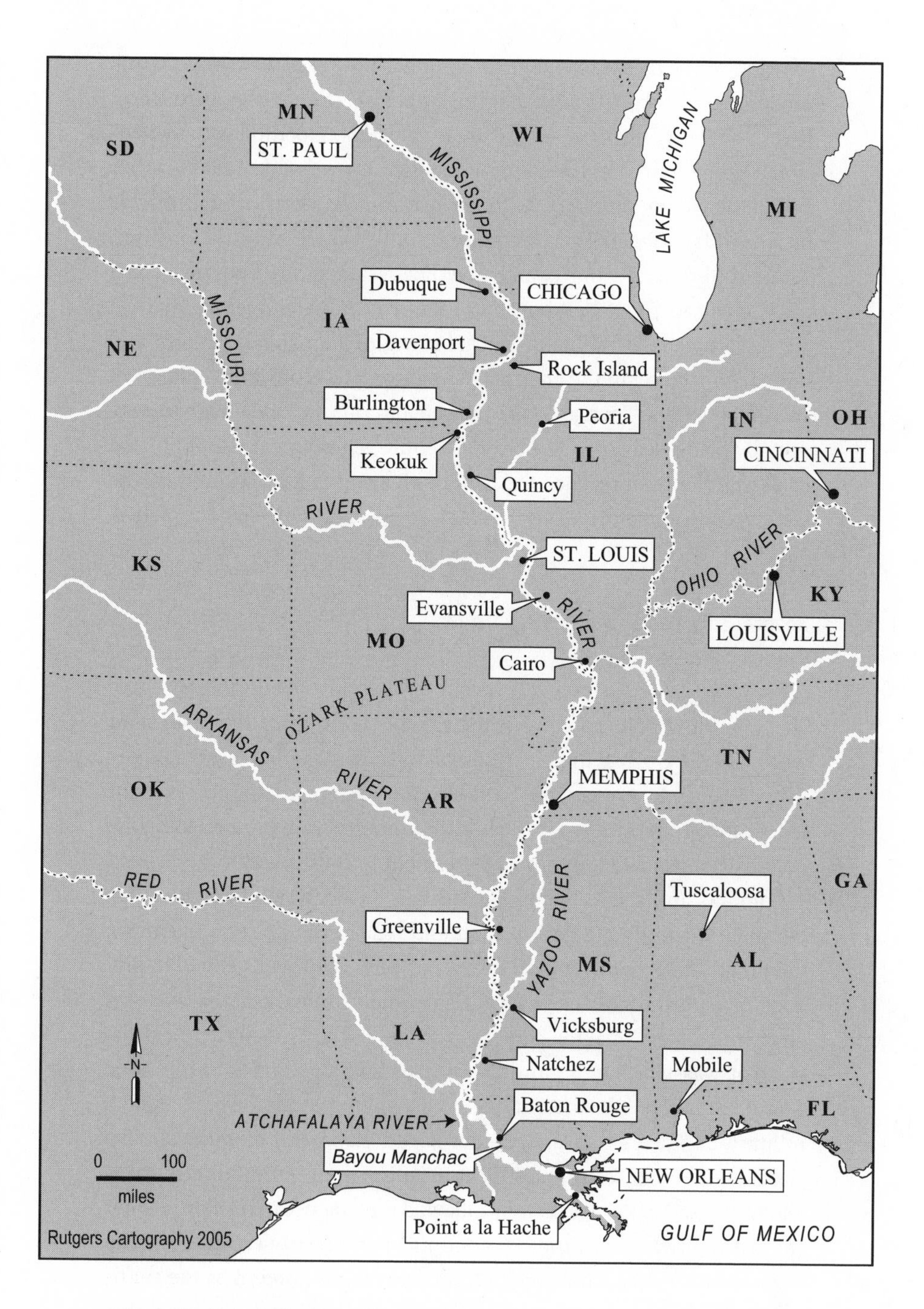

MAP 1. Mississippi Valley

creating fertile delta areas. Before engineering works were built, it was moderately difficult to navigate the upper section of the Mississippi River, especially through several areas of rapids. Flooding was an occasional problem in many sections along the upper river.[2]

But the major problems with navigation and flooding, familiar to us from Mark Twain's accounts, were on the lower Mississippi River, south of Illinois. During Twain's riverboat years, pirates, obstacles, and tricky currents made that portion of the river more dangerous than an ocean voyage.[3] Caving banks caused great, hidden shoals to form and reform constantly in the channel, demanding the constant vigilance of steamboat pilots. Then as today, the last 450 miles of the river bed were actually below sea level. The river's volume was so great that it pushed itself uphill until it reached the Gulf of Mexico.[4] Political claimants and settlers confronted these physical features as conditions to be managed.

Sovereignty and Development under Colonial Authorities

To contending political powers, the value of controlling the lower river was evident. During their periods of rule over the lower Mississippi River, France, Spain, and the United States each regulated access to the river and to riverfront lands. They built infrastructure projects to secure political control and to promote settlement by loyalists.

Colonial authorities who claimed the lower Mississippi River attempted to coordinate government and landowner projects to control flooding. The original French settlers of New Orleans intended to produce agricultural surpluses. They also vaguely understood that swamps were sources of "miasma" that caused fever. The French therefore decided it was worthwhile to drain nearby swamps and to build flood works, even before laying out city streets and building the port. French military engineers had already developed elaborate, scientifically derived engineering techniques for controlling their home country's major rivers.[5] In New Orleans, they followed the practice of residents along silt-bearing rivers in Europe by building artificial earthen levees on top of the natural levees that had built up from slight floods over the years.

French colonial authorities began building a levee at the New Orleans town site in 1717. They granted rural riverfront lands to planters on the condition that they build levees on these properties.[6] Police juries could compel those who owned land within seven miles of the river to provide slaves to build and maintain levees. This power was especially important during flood emergencies.[7] Some landowners organized informal cooperative efforts to build neighborhood levees. But because the artificial levees at that time were only three feet high, the best hope was that a levee would break a safe distance upstream to reduce the river's volume past one's own plot. By 1735, slaveholding cotton and sugar planters had built levees on both sides of the Mississippi River, extending 30 miles north and 12 miles south of New Orleans.[8] Spanish colonial authorities, who took possession of the province of Louisiana in 1763, extended similar requirements to their land grant recipients.[9] Navigation and flood control works produced through private and public effort were therefore characteristic of political rule in the lower river region from the very start of European settlement.

The Louisiana Purchase, the First New U.S. Territory

New Orleans became strategically vital by the mid–eighteenth century, when French and British established rival forts on tributaries in the Northwest Territory, and when settlers from the Atlantic colonies began moving west in great numbers. European powers characteristically set out plans for settler colonization in the West and then imported settlers. By contrast, the U.S. government was pressed to seek control of the lower Mississippi River because settlers from the United States had moved west and began to stir trouble. It was only when the strategic aims of France and Spain changed in the late eighteenth century that the extraordinary Louisiana Purchase became possible.

Control over the lower Mississippi was at the heart of every foreign rivalry that the United States experienced in that region. When the Atlantic seaboard colonies declared independence from Britain in 1776, they claimed territories west to the Mississippi. Despite these claims, British and Spanish colonial authorities and freelance toll companies limited settlers' access to the Mississippi River.[10]

Western settlers began to demand that the U.S. government earn their allegiance by guaranteeing their right to free transit on the Ohio and Mississippi rivers. They expressed their demands as clear statements of rivalry with the eastern states. In 1783, England recognized U.S. independence and agreed that both nations would perpetually enjoy free navigation on these rivers. Fearing economic competition from the United States, the Spanish refused to recognize this agreement. From 1782 to 1794, Spain continued to charge settlers from the United States fees for using the Mississippi.[11] Leaders of northeastern states who identified their economic interests as distinctive from those of the West openly expressed their willingness to sacrifice settler access to the Mississippi River, in the hopes of securing reciprocal trade with Spain.[12] In response, settlers in Kentucky organized Democratic Societies, modeled after Jacobin clubs. They urged using all available methods to guarantee free navigation, including a plan that Citizen Genet of France proposed to Jefferson to raise an army in Kentucky and seize Louisiana from Spain.[13]

With its empire failing and the burdens of supporting the Louisiana colony exceeding its worth, Spain's king recognized settlers' navigation rights and abandoned Spanish posts on the east bank of the Mississippi in 1795. He then retroceded Louisiana to the French in 1800.[14] Word of the deal alarmed politicians and settlers in the United States. When the caretaker Spanish governor of Louisiana rescinded U.S. citizens' right of transit at New Orleans, their alarm increased. Federalists demanded war. Jefferson resisted, feeling that conflict with France would require the United States to ally with England. Jefferson then sent James Monroe to Paris in 1803 to propose buying New Orleans and West Florida. But because he was losing the war in Haiti, Napoleon had already lost interest in building a North American empire. Even before Monroe arrived, France's foreign minister, Charles Talleyrand, offered all of the Louisiana territory to the resident U.S. envoy. Astonished U.S. diplomats accepted that offer within the week. Jefferson later approved it, despite there being no provision in the Constitution for acquiring territory.[15]

River Defenses and the Beginning of Army Engineer Involvement

The U.S. Army Corps of Engineers soon took over French posts in New Orleans and elsewhere in the Mississippi Valley. The United States had already built forts at the present-day sites of Vicksburg, Chicago, and elsewhere. Soldiers at the forts were sentinels and aggressors against British troops and native peoples living near the upper Mississippi River; the Cherokees, Choctaw, and Chickasaw near the lower Mississippi River; and British, French, and Spanish ships in the Gulf.

Continued conflicts with the British prompted Congress to fund some improvements of coastal defenses and harbors from 1807 to 1812, including defenses at New Orleans. During the War of 1812, permanent and improvised defenses failed to keep the British from entering waterways near New Orleans.[16] In January 1815, however, Andrew Jackson's forces defeated the British as they attempted to cut a levee and advance against U.S. troops.[17]

Sovereignty in the valley was not yet settled. The United States took Florida in 1819, although tensions with Britain continued for some time. Most importantly, the forced displacement of nearly all native peoples east of the Mississippi between 1820 and 1840 opened the lower valley for a new cotton economy based on slavery. Improvement of the lower Mississippi River became a pressing matter to residents and investors.[18]

Putting Resources to Productive Use under U.S. Rule

After the 1812 war, thousands from all parts of the United States moved to the new territories to buy land speculatively or to farm. Some settlers aspired to become small farmers. Others came as wealthy slave owners, moving with their slaves from declining southeastern lands. These planters established sugar plantations in southern Louisiana state or corn and cotton plantations in the lower Missouri River valley and along the west bank of the Mississippi River near St. Louis.[19] The Mississippi river valley had 1.4 million white settlers by 1810 and 2.6 million by 1820.[20]

The slave plantation system was firmly established in the lower Mississippi river valley by the 1830s. The new plantations revived slavery as a labor system in North America and increased shipments of cotton and sugar down the river to the East Coast, England, and France.[21] Rural land concentration around New Orleans was already high because the U.S. government had agreed to recognize land titles granted by the colonial French and Spanish governments. The new plantations increased land concentration further.[22] Cotton arriving from Alabama, Mississippi, Louisiana, and Texas made the port city of New Orleans the third largest city in the country by 1840.[23]

As landowners extended cropping along the Mississippi River, they extended levees. States along the Mississippi River organized ad hoc cooperative flood control efforts in the early years. In 1803, the most substantial levees ran only as far north as Baton Rouge on the east bank and a bit farther north on the west bank.[24] In its first bill concerning levees, the Mississippi legislature created a commission in 1819 to assess up to $8,000 from landowners near Warrenton, south of Vicksburg, to build levees.[25] Bayou Manchac was closed in 1828, the first in a series of closures that increased the flow on the river's main stem. Levees in southern Louisiana were continuous, if small, on both sides of the river as early as 1828, but crevasses and floods were frequent.[26]

The costs of cooperation and of the failure to cooperate were already apparent in the early nineteenth century, to judge from stories that were told. Those suffering flooding from a crevasse often asserted that a neighbor had cut the levee to forestall flooding on their own lands. Mark Twain reported another tactic:

> When the river is rising fast, some scoundrel whose plantation is back in the country, and therefore of inferior value, has only to watch his chance, cut a little gutter across the narrow neck of land [of a horseshoe curve] some dark night, and turn the water into it, and in a wonderfully short time a miracle has happened: to wit, the whole Mississippi has taken possession of that little ditch, and placed the countryman's plantation on its bank (quadrupling its value), and that other party's formerly valuable plantation finds itself away out yonder on a big island.[27]

Reports of such incidents over the years were exaggerated, conditions having to be just right to cut a meander or a levee. Even so, ensuring levee safety was second only to controlling slaves as a security concern for planters. Individual planters and levee district officials often took the precaution of patrolling their levees during times of high water to prevent sabotage.

Navigation Improvements as a Key to Western Development

While the immediate inspiration for Congress's 1824 navigation improvement legislation was the court dispute over New York's waterway monopolies (see chapter 2), demand for aid from the Mississippi Valley was politically critical for passing the bill. All lands along the Ohio and Mississippi rivers were organized into states by 1819. Members of Congress from these new states secured public lands within their states to finance roads, canals, and other infrastructure projects.[28] Steamboat companies on the Ohio and Mississippi rivers offered regularly scheduled passenger and freight service by 1817, displacing the small keel boats in use since the mid–eighteenth century.[29] While dozens of steamboats exploded dramatically and bloodily, especially during the early, unregulated years, many more were damaged or sunk by tree branches or shoals in the channel.[30] Legislators began to see federal river navigation improvements as a key to regional growth. In the first navigation aid bill in 1824, Congress provided $75,000 to Corps units in New Orleans to remove snags on the Mississippi and Ohio rivers and to experiment on reducing sand bars. In 1837, Congress also authorized channel dredging on the Mississippi.[31]

As the western economy grew, residents began to draw up grand plans to develop the Ohio-Mississippi river system. Some plans centered on railroad development, some on river development, and some on a combination of the two. Mississippi River activists were optimistic. A Corps engineer in 1839 estimated that shipping a ton of goods on the Ohio-Mississippi river system would cost as little as a half cent per mile, as compared to two and a half cents on the primitive rail systems of the time.[32]

Northeastern merchants meanwhile proposed rival transportation networks to guide southern and midwestern goods through eastern markets. By 1830, the northeastern states had developed their industrial bases to the point where they no longer depended on the trade in southern cotton.[33] Traffic peaked on the Mississippi River in the 1830s and then declined due to banditry, excessive port fees, and the failure of shippers to provide reliable service. New rail lines built in the 1850s from the East to Chicago and other midwestern cities drew away more business. On some lines, railroads manipulated rates to undercut river transportation until local river shippers went out of business. But the railroads only hastened a decline in river traffic that was already in progress.[34]

The Social Organization of Environmental Change Takes Shape

The outlines of the twentieth-century riverfront flood control system were drawn before the Civil War. Levee districts were organized by states or local governments to supplement private levee building in the 1830s and 1840s. Landowners began by forming levee boards to coordinate flood control in a given area. They then demanded from their state governments the powers to police and tax their fellow local landowners. Taxes usually focused on the riverfront landholders because most colonial settlements along the river had been only one-plot deep. Despite planters' complaints, it took decades for states to tax backlands owners who also benefited from levees.[35]

State levee legislation typically required each county board to establish several levee districts and to appoint inspectors for each district. Inspectors would plan levees and drainage ditches, would inform each riverfront landowner what work would be done, and could fine levee inspectors or landowners for neglecting their duties. When water threatened the levees, inspectors could order out the slaves of delinquent planters.[36]

By the 1840s, even though most levee districts lacked the population to finance complete levee systems, the east bank levee line extended north from the New Orleans area nearly to the Arkansas River in southern Arkansas. East bank levees had many gaps along the Yazoo-Mississippi Delta, the area between the Yazoo and Mississippi rivers in Mississippi state. In the Delta, a few planters recently arrived from old cotton areas had directed their slaves to clear strips of the jungle for plantations on the natural levees just along the river.[37] Levee quality in Louisiana was often poor, as laws setting levee size, shape, and distance from the river were not enforced. Even some of the "best" plantations there lacked full levees.[38] As more levees were built, however, the danger of a levee break increased, because the levees allowed the river to carry a greater volume of water.[39]

Many prewar projects were delayed or blocked or otherwise defied engineers' best predictions. For example, plantations with levees were flooded due to a neighbor's negligence. Those without levees were flooded when downstream levees pushed high waters back upstream.[40] Meanwhile, channel clearing ordered by the 1824 Rivers and Harbors Act was delayed. The first contractor interpreted his contract very narrowly and did little work, and the government had him in court for at least forty years. A $285,000 dredging operation to deepen the river's passage to the sea was lost when the entire river shifted to another channel.[41]

The most serious physical problems resulted from two projects that pilots and merchants promoted and that the public celebrated. These projects were intended to shorten the shipping route to New Orleans and to provide better access from the Mississippi west and south through Louisiana's Atchafalaya swamp. Steamboat inventor Henry Shreve directed both projects as the Corps' supervisor of navigation contractors in the 1820s and 1830s. Shreve first cleared the 150-mile raft of logs on the Red River near Vicksburg with his new snag boat. He then cut off a nearby meander of the Mississippi River. Shreve had so much popular support that President Andrew Jackson approved about $3 million for river clearing, more than any other prewar president spent on rivers.[42] But more than any other interventions, these two projects inadvertently worsened flooding and navigation conditions over the next century by hastening the Mississippi's flow into the swamp and away from its current main channel past New Orleans.[43] Thus began a pattern that is now familiar in environmental programs. The government became increasingly committed to managing river conditions, in part to control the unintended consequences of its earlier interventions.

Conclusion

Extending U.S. rule in the lower Mississippi Valley affected the way that the central government would relate to all regions. It committed the government to a policy of territorial expansion. It also brought on the difficulties of integrating a culturally and economically distinctive region.

In the lower Mississippi Valley, the most lucrative opportunities for expanding plantations appeared to be the rich but hazardous bottomland areas. Planters and shippers began to envision federal aid for navigation and flood control as a means for expanding the economic base of cotton and sugar agriculture. With this base, they aimed to establish regional control over the cotton trade with England. Legislators from all regions eagerly took up the practice of logrolling after the Corps of Engineers' river navigation program was created in 1824. Legislators from the southern portions of the Mississippi Valley in particular refined the practice of expressing regional resentment by making claims for resources as they tried to influence appropriations in river navigation bills. Few demanded federal flood control aid before the 1840s, but landowners had by then created local and state institutions supporting levee building.

By the 1830s, planters and merchants who dominated the emerging cotton and sugar economy of the lower Mississippi had defined their own region's interests as distinct from those of the northeast even while arguing that navigation aid to their region was a national duty. They began to organize a campaign for river aid that they felt would increase regional independence while relying on national revenues for support. To many of these southern leaders, further U.S. expansion into far western territories appeared to threaten the South's political and economic standing. Leaders proposed a variety of responses. These included blocking the admission of new nonslave states, improving shipping channels from the upper Middle West through St. Louis and New Orleans, and shifting control over the cotton economy from New York to New Orleans. By accentuating the many differences between North and South, such proposals heightened the rivalry between states.

chapter 4

The Mississippi River

RESENTMENT LEADING TO CIVIL WAR

Southern Growth Slows, and Regional Relations Change

Southern ambitions for economic autonomy stalled by the 1840s, and New York retained its financial control over the cotton trade. More and more of the grain grown in the booming upper Middle West was shipped via rail and waterway through Chicago to the East, bypassing St. Louis and New Orleans. This shift, plus the spread of abolition in the North and the admission of nonslave states in the Far West, left the South not only physically isolated and economically underdeveloped, but also socially distinctive and politically resentful. The disputes leading to the Civil War are familiar. The special contributions of infrastructure politics to war tensions are recounted in this chapter to highlight processes of territorial expansion and integration that contemporary leaders experienced as central to the conflict. Prewar regional competition for development aid emerged after the war in new form, with Mississippi Valley navigation and flood control advocates arguing that aid would rebuild war damage and help reintegrate the South into the national economy.

Decisions to organize western territories as states, rather than colonies, meant that decisions about property rights, settlement, and land development would shape the practice of federalism. The Northeast relied on the southern cotton trade for decades, but once the Northeast's industrial base was built, politicians there were less inclined to support improvements to the Mississippi River. National control over overseas trade had been a vital issue for those seeking independence. By the 1830s it was clear that domestic transportation links would determine which region of the United States would benefit most from international as well as domestic trade.

The federal government did not determine the regional winner outright but set conditions that encouraged the regions to compete over policy and the market. By the late 1850s, the mainland territories had been acquired. Government policy for internal improvements and settlement was expressed through a patchwork of activities. These included land grants as well as land sales, technical consulting by army engineers, and a range of private, public, and combined ventures that the federal and state governments encouraged or sponsored. The army in the United States had not taken the lead in designing and building an integrated national system of infrastructure, as it had in France.[1] Even so, the U.S. federal government had encouraged growth, engineered some rivers, and given land grants or technical aid to most of the major railroad networks. State and local governments and private investors were effectively part of the emerging system of infrastructure. Development of the country's vast territory depended largely on the ability of each region's leaders to use state and federal government policies to attract settlement and private investment that suited the region's economy and social base. The path favored by most politicians in the lower Mississippi Valley was to promote export-oriented cash cropping by slaves. This path did not lead to sustained growth. Instead, it became an obstacle to national unity.

Advocacy Begins: River Improvement Conventions

Internal improvements became entangled more deeply in sectional politics in the 1840s. Advocates of river aid who were frustrated with small appropriations for the Mississippi Valley began to organize lectures, meetings, and conventions to attract attention to their cause. Most of the conventions calling for river development in the Mississippi Valley were organized through city-based commercial and civic organizations. Over time, conventions increasingly became expressions of sectional interests and identity.

Advocacy conventions were a part of a new political "repertoire"[2] for addressing cultural and economic development. Improved transportation networks allowed people to travel from greater distances to attend meetings. Coupled with newspaper coverage and advertisements, conventions and lectures became the means for communicating concerns directly to politically active members of the public. Conventions also became important stops for politicians, some of whom communicated convention resolutions to Washington. Members of the rising middle and upper classes called conventions to advocate for temperance, railroad building, abolition, or river development. Commercial organizations and local government agencies that called river improvement conventions (see appendix 1) were most vigorous in the Middle West and the South. Activists there argued that Congress had been disproportionately generous in granting river aid to northeastern rivers and to railroad developers.[3]

Commercial capitalists integrated river improvement campaigns into their general strategies for regional growth. For example, citizens and merchant groups in St. Louis set special meetings to discuss Mississippi River improvement in the 1840s, even before St. Louis was much of a city.[4] "Citizens" and the City Council of Cincinnati sent a memorial to Congress in 1844 urging development of the Ohio and Mississippi rivers.[5] In 1846, the Mississippi legislature sent the first of many memorials they would write for flood control aid. This one requested that the national government grant the state specific parcels of swamplands where state officials wished to build levees.[6] The Chamber of Commerce of the state of New York sent a memorial to the president in 1859 requesting that the federal government improve the Mississippi River to improve commerce for the entire nation.[7]

From the 1840s to the 1910s, landowners, levee district managers, politicians, shippers, and businessmen in the Middle West and South attended river improvement conventions in port cities connected to the Ohio-Mississippi River system, including Memphis, Dubuque, St. Louis, Vicksburg, New Orleans, and Chicago. Nearly any convention could attract local congressional representatives. Many attracted governors and senators from the region. Between meetings, convention executive committees and their sponsoring organizations sent memorials to state and national legislatures and published pamphlets extolling public action.

Debate in Congress, in newspapers, and in conventions often concerned constitutional justifications for river improvement. The interstate commerce clause had justified navigation aid in 1824, and it was most frequently mentioned. In the memorial to Congress sent by the convention held in Chicago in 1847, delegates meticulously cited the commerce clause, the power of Congress to dispose and regulate territory and property belonging to the United States, the power to collect taxes and pay debts, the common defense clause, and the general welfare clause.[8]

Activists from the Ohio River and the upper Mississippi River typically demanded channel and harbor improvements, while those from the lower Mississippi River demanded flood control as well. Southern activists argued that the difficulties of maintaining flood control levees and shipping channels on the lower river had to be solved in tandem because they had a single cause: a river that carried large volumes of silt and rapidly shifted its banks. The collected floodwaters of half the nation burdened the South by worsening these conditions, they argued. Furthermore, states along the lower river found it difficult to maintain bridges and levees that were used by trains involved in interstate commerce. It was, therefore, clearly a duty of the central government to facilitate commerce. When veterans of the Mexican War began buying parcels on the Mississippi River in 1849, activists cited support for veterans as yet another reason why the federal government should provide flood control aid.[9] Other convention speakers countered that they had trouble enough winning navigation aid. Conventions held before the Civil War seldom even considered passing resolutions calling for flood control aid.

Development and Sectionalism

Sectionalism dogged river improvement activism. In the 1840s and 1850s, activists and legislators from the Northeast, upper Middle West, and South proposed incompatible schemes to link the new agricultural area of the upper Mississippi valley by river or railroad to international ports in different cities.[10] Democrat John C. Calhoun of South Carolina had supported Henry Clay's nation-building program of public works in the 1810s. He became a living symbol of southern contrariness by the 1830s, after splitting with President Andrew Jackson and leaving the vice presidency. Back in South Carolina, Calhoun became a leader of the states-rights movement, which aimed above all to keep the federal government from interfering with slavery.[11] Calhoun organized the Mississippi River Improvement Convention in Memphis in 1845, which called for a canal connecting the Great Lakes to the Mississippi River and for improvements on the Mississippi and other western rivers. Rail lines would also link the Mississippi River to Savannah and to the port cities of Charleston, Richmond, and Baltimore. Calhoun included many of these propositions in a 1846 Rivers and Harbors appropriations bill, which Democratic President James Polk vetoed, condemning the "disreputable scramble" for aid. Despite his hope in attracting support, Calhoun's plan to build southern infrastructure alienated many legislators from the Middle West.[12]

As an indication of midwestern sentiment, the largest river convention met in Chicago in 1847 to protest both Calhoun's attempt to draw commerce south and Polk's veto of a 1846 river and harbor bill. Andrew Jackson had favored Polk for president. River aid supporters had therefore expected Polk to follow Jackson in supporting western river funding. Instead, Polk provoked their ire by actively and publicly opposing river projects.[13] William M. Hall, an agent of the Lake Steamboat Association, formed the committee to plan the Chicago convention. He called on political officials and leaders of commercial organizations in Chicago, Detroit, Cleveland, Buffalo, Syracuse, Rochester, Utica, Albany, Hartford, New Haven, Boston, and Providence, firmly snubbing the South. A. B. Chambers, a member of the convention's General Committee who worked with the St. Louis Chamber of Commerce, was unable to persuade Hall to move the meeting from Chicago to St. Louis. Organizers called preconvention meetings with press briefings in several cities in order to gain local press coverage.[14] The convention itself gained national publicity for attracting over 10,000 delegates to the city of 16,000.

Almost all delegates were from the Northeast and upper Middle West. Prominent delegates included Horace Greeley, Liberal Republican and Democratic presidential candidate in 1872; Schuyler Colfax, later vice president under President Ulysses Grant; and the new Illinois representative to Congress, Abraham Lincoln.[15] Greeley wrote daily newspaper dispatches. Convention delegates passed resolutions supporting river improvement and the building of transcontinental rail lines that would link the Far West to the rivers of the upper Middle West, not the South.[16]

Despite William Hall's aim of organizing a nonpartisan convention, the convention occasioned criticism in Democratic newspapers and praise in the Whig press. Hall later complained that Whig delegates in particular presented convention reports that glorified the Whigs and downplayed participation by Democrats.[17] One Whig magazine attacked Lewis Cass, the Democratic nominee for president in 1848, because he skipped the convention.[18] Similarly minded persons issued a two-inch-by-one-inch pamphlet of Cass's letter of regret to the convention with an "errata" changing his name to "ass."[19] Cass, who had been secretary of war under Andrew Jackson, no doubt avoided the convention for fear of alienating southern Democrats.[20] As a further partisan tactic, speakers cited statistics and statements in support of government river projects that had been published by President Polk's own chief of the Corps of Topographical Engineers, the Whiggish John J. Abert.[21]

Even as tension increased between northern and southern activists, other conventions continued to promote unified development of the Mississippi Valley as a way of forestalling domination by northeastern businesses. The 1846 Memphis convention drew delegates from upper and lower Mississippi River areas. Local conventions—such as those in Evansville, Indiana, and Burlington, Iowa—also promoted improvements for the entire length of the river.[22] These few statements of river unity determined neither congressional policy nor consistent cooperation among activists from the North and South.

When Polk vetoed a small rivers and harbors bill in 1847—saying that internal improvement projects would interfere with spending for the Mexican War—Congress passed resolutions denouncing Polk's veto. Congressional power to fund public works was an important issue in the 1848 presidential election, even though the winner, the Whig Zachary Taylor, avoided commitment on the issue.[23] Members of Congress responded to the controversy with a characteristic gesture, by giving away land, instead of aid, to promote localized development.

Swamp Land Acts

After years of agitation by river activists and a new series of floods on the Mississippi River, Congress granted all "swamp and overflowed lands" within Louisiana's borders to the state government so that it could finance flood control works.[24] Convention delegates, newspaper editors, and politicians had argued over the years that the federal government should grant unused wetlands to states for levee building.

Politicians from Louisiana—where nearly one-third of the land was swampy, and even malarial—had especially favored a grant program.[25] Congress considered this proposal several times from 1826 on but didn't pass the first Swamp Land Act until 1849,[26] after the dispute with Polk had raised members' interest in asserting their congressional will. Few members opposed the bill. General wisdom held that the lands would have brought little money to the U.S. Treasury if sold through regular government land sales. And northeastern politicians may have acceded to the grant program in the attempt to keep leaders of the upper Mississippi River Valley from allying more closely with leaders of the lower valley.[27]

When river and farming interests from other public land states demanded similar grants, Congress extended the program to California, Florida, Oregon, and eleven states in the Ohio-Mississippi Valley.[28] Congress also approved an army survey of the lower Mississippi River in 1850 to discern the river's mechanisms and to recommend improvements to navigation and flood control.[29] Nearly 65 million acres passed into private hands through the swampland program, some for as little as $.10 to $1.25 per acre. Sixty percent of the grant acreage was concentrated in Florida (20 million acres), Louisiana (over 9 million acres), and Arkansas (over 7 million acres).[30]

The program was beset by lawsuits. Complaints about corrupt state-level deals and about land speculation provoked repeated government investigations. In 1915, a Commissioner of the General Land Office—which administered the program—concluded that recipient states had failed to promote drainage of most of their swamplands. He judged the program to be the greatest failure of the federal government's land grant programs.[31] Land claimed as swamp included parcels in the Ozark mountains. Plots of other supposed swamps were adjacent to lands that had been claimed under the desert land grant program. According to public land historian Paul Gates, the primary problem was the vagueness of the acts themselves. In particular, the swampland acts failed to require states to actually drain land or build levees in order to gain title.[32]

The grants had several unintended consequences, including acceleration of the growth of market-oriented agriculture and forestry. On the lower Mississippi River, most of the grants passed to absentee landowners, and most of the absentees were Yankees.[33] Several large southern lumber companies got their start by claiming and logging swamplands that were in reality thick with cypress trees or long leaf pines.[34] A good proportion of the swamplands in the Yazoo-Mississippi Delta were sold under this program.[35] In examining swampland disposal records for Indiana, Illinois, Louisiana, and California, Paul Gates concludes that the majority of the parcels went to people already involved in large land investments.[36]

The swampland grants also impelled flood control activism by increasing the ranks of market-oriented landowners who owned flood-prone lands. Some of the areas that received swampland grants experienced river flooding as the primary obstacle to agricultural development, namely parts of Louisiana, Mississippi, and Arkansas near the Mississippi River and its tributaries, and lowlands near California's Sacramento River. These were the areas that first produced flood control activism. Owners in these regions organized new levee districts, promoted the flood control cause in local and state politics, and demanded federal aid. Populists would later call these grant recipients "swamplanders."[37] In the other eleven states receiving swampland grants, river flood control remained a secondary issue. Landowners there could usually drain their lands without first having to build river levees.[38]

The swampland program also gave a boost to the flood control cause by sponsoring Corps surveys. These surveys consistently recommended that the federal government build protective projects.[39] And new levee districts organized by swamplanders became local contacts for the Corps along stream reaches where Corps navigation work affected levee building.[40] So the swampland acts involved Corps engineers in assessing the viability of grant lands. The acts also unintentionally created an interest group by reinforcing the dominance of capitalist largeholders in the valley, educating them to gain favors from the central government and states, and organizing landowners politically through the creation of additional levee districts.

Internal Improvements, North versus South

Advocates of the military used war, territorial security, and disasters to justify funding further civilian work by the army. Congress and the president had cut funding for army engineering projects in the 1830s and 1840s. Embezzlement scandals, a patent dispute by Henry Shreve, and the failure of army projects on several rivers had soured enthusiasm for navigation work. But quick victory in the Mexican War, acquisition of Mexican lands, and flooding of 27 million acres along the Mississippi led to a new batch of civil works. These began with $50,000 for a survey of the lower Mississippi in 1850 and a $2 million appropriation for rivers and harbors in 1852.[41] Under President Millard Fillmore's direction (1850–53), the army split the Mississippi survey money to fund two competing surveys. One, completed by Charles Ellet in 1852, called for a complex series of high dams and floodwater reservoirs that would release excess water during low water season.

A more detailed and much more influential survey by A. A. Humphreys and E. H. Abbot was released after the Civil War. Their report concluded that building continuous levee lines would "concentrate" the flow of the river and scour the channel. They rejected digging cutoffs to bypass bends in the river, and they recommended closing the remaining natural outlets that drained water away from the main channel and into swamps. This single-channel theory tied flood control needs to navigators' needs for a deep channel. It was, therefore, an important step for an agency that had devoted itself solely to navigation improvement work. Humphreys became chief engineer in 1866, and the approaches outlined in the Humphreys and Abbot report became standard ideology for the Corps of Engineers.[42] Abbot generated public support for the report by publishing a review of it in the *North American Review* and by circulating it as a pamphlet.[43]

Many politicians in the South continued to oppose internal improvements on principle. Others argued that transportation links between the South, the upper Middle West, and the Far West could enable New Orleans to compete with New York City for trade in midwestern grains and other goods.[44] As president, the Democratic southerner Franklin Pierce (1853–57) and Secretary of War Jefferson Davis promoted army surveys and civil works in the 1850s, including the southern transcontinental railroad line.[45] But illegal river tolls, local shipping monopoly fees, and severe competition in river sections not already monopolized had already severely cut traffic on the lower Mississippi River. Rail connections between the East and the upper Midwest had helped to shift trade away from the South.[46]

Levee Districts Expand

While the swampland grant program provided little funding for flood control works over the years, it did stimulate local organizing to improve flood control. Public land historians often dismiss the program, saying it failed to improve flood control. Indeed, speculators who received many of the parcels held them unimproved for years, and most state governments redirected some of the land or land sale funds toward other government activities.[47] On the other hand, historians of local reclamation programs report that the grants did finance levee lines in California, Arkansas, Louisiana, and Mississippi. These were precisely the regions that became centers for flood control activism.[48]

For example, newspaper editors and bottomland property owners in Mississippi encouraged levee building under the program to attract planters from states bordering the Mason-Dixon line, where movements to abolish slavery were growing. Many of these planters did move to the forbidding and sparsely settled Yazoo-Mississippi Delta during the 1850s.[49] In 1858, *De Bow's Commercial Review* announced, "If both sides of the river cannot be leveed, then we [Mississippi state] must protect ourselves, and let our neighbors in Arkansas suffer."[50]

Arkansas, Mississippi, Missouri, Kentucky, Tennessee, and Louisiana passed levee district laws in the 1850s in response to the swampland grants. Louisiana and Arkansas created state agencies to direct most of the grant program's levee-building work. The local importance of levees was reflected in Mississippi laws that made levee cutting during high water punishable by three to thirty years in prison for whites and by death for free and slave "negroes" and "mulattos." Projects financed from swampland sales linked existing levees built by planters and levee districts. Most of the major lower valley basins had nearly complete lines of small levees by the mid-1850s.[51]

Planters used slave labor to build levees and to dig drainage ditches in newer settlements. As cotton and sugar plantations expanded, however, the cost of slaves rose. Planters decided that slaves were too valuable to risk doing heavy, often lethal labor in the malarial wetlands. Planters instead engaged labor contractors to hire Irish and other immigrant laborers directly from ships arriving from Europe in New Orleans, Cairo, St. Louis, and Cincinnati.[52]

As state governments became more involved in levee regulation and finance, governance became more complicated. Mississippi state required levees to be set far back from the riverbank in the 1850s. In response, some owners of unprotected lands next to the river sought damages as a condition of granting the rights-of-way to build levees.[53] In Mississippi state, early levee financing laws pushed many landowners into foreclosure.[54] Although riverfront landowners continued to pay most levee costs, railway companies built some levees. Local and state government programs also increasingly socialized the costs to owners of backlands who also benefited from riverfront levee lines. By the late 1850s, many districts in Missouri, Arkansas, Mississippi, and Louisiana taxed all alluvial landowners, providing a base for increasingly professionalized infrastructure programs.[55]

At the time of the Civil War, Mississippi River floodplains in Arkansas, Louisiana, and Mississippi had the highest concentrations of slaves and the largest plantations and were among the most important cotton areas. Riverside planters also grew rice, and in southern Louisiana, sugar.[56] West bank levees had many gaps south of Cairo, Illinois.[57] East bank levees were complete from Memphis to Vicksburg. Although Mississippi River levees were still low and weak along much of the Yazoo-Mississippi Delta, a new levee board there had completed a number of levees.[58] There were few levees between Vicksburg and Baton Rouge, but levees were complete between Baton Rouge and Pointe a la Hache, south of New Orleans. According to a comprehensive Corps survey of the Mississippi River, levee lines reached their greatest prewar extent from 1850 to 1858.[59]

Defense, Rivers, and the Civil War

Regional conflicts over land and infrastructure development only added to tensions leading to the Civil War. Planters debated whether the far western territories would be climatically suitable for slaves. Southern politicians worked to prevent an increase in the number of nonslave states. Congressional debates in the 1850s concerning the admission of slave and nonslave states were ominous. In response to belligerent southern talk, Henry Clay recalled Spanish and French efforts to cut off shipping access to the lower Mississippi. Clay asserted that residents who had since settled upstream would not allow the southern states to once again make the river's mouth foreign territory.[60] After benefiting from the quick growth of the cotton economy and New Orleans trade, southern leaders saw these gains threatened by the abolition movement, by the growth of Chicago as a trade hub, and by a series of economic development bills that would benefit the Northeast.

In the 1850s, as Chicago was growing, New Orleans and other southern ports had enjoyed a steady business shipping goods to the Northeast and to Europe. At that time, it was cheaper to ship goods to Liverpool from New Orleans than from New York.[61] But Chicago soon became the center for middle western grain exports. In Chicago, eleven major railroad lines to eastern ports were built with the aid of federal government surveys, private consulting by army engineers, federal tariff relief on iron imports, and federal and state land grants. The rail lines also gave Chicago an advantage over St. Louis and New Orleans in the bid to build the first midwestern rail connection to the Pacific coast.[62] Southern leaders began to recognize that many road, water, and rail links between the East, Midwest, and Far West had already been built or planned.[63]

Then, in the late 1850s, northern Republicans introduced a set of bills that would affirm the dominance of northeastern businesses in trade to the Far West.[64] Most southern politicians at that time, whether former Whigs or not, favored or tolerated central government internal improvement projects. Most also opposed the benefits that this particular set of bills would target to the North. Plantation slavery had thrived in the 1830s because northeastern banking and shipping firms benefited from selling southern produce. The new set of bills concerning tariffs, taxes, banking, subsidies, and land grants would boost the Far West as an agricultural region in competition with the South. It would also increase links between the Northeast and the Far West. Reading these proposed bills, southern politicians saw that the impending Republican control of the federal government could permanently subordinate the cotton South and doom southern leaders' hopes of establishing a direct cotton trade between New Orleans and England.[65]

Congress began passing many of these improvement bills even as southern states seceded.[66] The Far West, rather than the South, had become the key to the continued economic growth of the Northeast. This new regional interdependence blocked the possibility of a reactionary alliance of northern industrialists with southern planters, such as occurred in Germany.[67] Southern states organized the Confederate States of America on 8 February 1861. As Trotsky remarked, "By its very nature such a state of affairs cannot be stable. . . . The splitting of sovereignty foretells nothing less than a civil war."[68]

Conclusion

Market-oriented southern planters had pressed for policies to sustain capitalist commodity production by slaves. By the 1850s, support for these policies outside the South had dropped, and politicians had hard work gaining appropriations for Mississippi River navigation aid. Northerners no longer accepted the idea that development aid to the South was of general public benefit, in effect questioning how the South related to the nation. Private investment and government development assistance by then had also built northern infrastructure networks to the Midwest. These transportation links reduced northerners' reliance on connections to southern rail lines and ports. Infrastructure politics increasingly became sectional politics, culminating in the series of major economic development bills in the 1850s benefiting the Northeast.

This marked the full emergence of a strategy for regional growth that would become characteristic of the federal system of the United States. Regional boosters now considered it a top priority to petition the federal government for aid to improve their region's competitiveness with other regions. Pork barrel politics had begun with the first river navigation bills. As opportunities for expansion to the Far West stimulated regional rivalries among existing states, southern leaders repeatedly complained about the amount of aid passing to the Northeast. Leaders from all regions openly discussed ways of using federal aid to their home region's advantage. Under the United States's federal system, the central government did not design national infrastructure systems. Instead, it set general policies about land and development, mostly in response to demands from the states and regions.

chapter 5

The Mississippi River

POSTWAR REUNIFICATION, POSTWAR AID

Postwar Politics Yields a New Form of Economic Intervention

Members of Congress quietly and informally approved flood control aid for the lower Mississippi River under the unsettled circumstances that followed Reconstruction. Even in the 1880s, few politicians or voters favored direct government aid for economic production, and most northern voters particularly opposed policies that would benefit southern planters. This chapter reviews events leading to the end of Reconstruction to explain why politicians from the lower Mississippi Valley pressed for this particular form of aid and why flood control aid became politically possible.

Infrastructure politics was taken up after the Civil War with an acute awareness of sectional differences. Politicians from all regions also understood that northeastern dominance was secure. Southern development after the war was hindered by war damage, disparities with the North, concerted competition from northern merchants, and the continued polarization of the Democratic and Republican parties. In contrast, rail, water, and trail links between the Northeast, Middle West, and parts of the Far West remained open. Much of the infrastructure in Union states was even improved during the war, partly because of prewar economic development bills passed by Republicans.[1] Southern elites returned to internal improvement politics with renewed certainty about the economic and symbolic importance of infrastructure for the South. They also brought a new set of arguments justifying special aid to their defeated and economically depressed region, focused on the nation's duty to all of its regions.

Few northern politicians acknowledged outright that they were appeasing southern elites by responding to these arguments, but leaders in the North and South concerned with maintaining peace often did so through decisions that amounted to appeasement. By the late 1870s, it had become clear that Reconstruction of the South was failing. Northern leaders aimed to end the federal government's direct supervision of the South, avoid new conflicts with southern elites, and maintain sufficient order to allow the business of business to proceed. Southern leaders aimed to secure an agricultural labor force, control the movement of freedmen, and revive the economy, all while reducing unfriendly federal oversight. The unraveling of Reconstruction provided an opportunity for southerners to press their demand for flood control aid as a way of meeting these aims.

In approving flood control assistance secretly, Congress made it even more likely that the Corps of Engineers would rely heavily on the existing resources of local institutions as it repaired and built levees. The Corps had always depended on state and local agencies and local contractors as it worked on navigation improvements because it had small field staffs and because project budget cuts could sometimes be made up with local funds. Corps engineers had restarted navigation improvements on the Mississippi River soon after the war. The Corps received no marked increase in funding for its new flood control duties, and so it could not build up a new staff to build levees. Landowners were eager to have the assistance. It was under these circumstances that the army began working closely with planters twenty years after the Civil War.

Political Development after the Civil War

War, reconstruction, and compromise realigned the central government's relations with the South and involved the government directly in land development. Social scientists attempting to understand government intervention during this period have focused on government regulation of the growing industrial economy.[2] But at a time when the majority of people still lived on farms, decisions about land and resources in courts and in Congress continued to set much of the direction for regulating property and for intervening into the wider economy.

The small central government of the postwar period would give rise to the imposing, interventionist government of the early twentieth century. In explaining this transition, some scholars have focused on the emergence in the late nineteenth century of active executive agencies. Stephen Skowronek argues that during the first half of the nineteenth century, courts and parties were concerned mainly with using procedural routines to maintain a balance between the central government and the regions. Party members sought federal offices in large part for the opportunity to distribute patronage and pork barrel projects, such as river navigation projects. Congress had little reason to build bureaucracies into organizations that might have acted to limit congressional discretion. According to Skowronek, it was only under the tremendous press of industrialization that the old patterns of "party domination, direct court supervision, and localistic orientations" began to break down in the 1870s, and new administrative powers began to develop.[3]

The Civil War did not directly build those bureaucracies. Both the Union government and the rebellious Confederate government centralized considerable power during the Civil War. But they dismantled nearly all of the administrative apparatuses of the war at war's end.[4] Likewise, the postwar program of pensions for Union soldiers and their families was not expanded into a general social program, even though it drew popular support.[5] But the war and postwar politics—including demands for southern river improvement—did broaden perceptions of what activities the government could undertake.

Wartime interventions into the economy had been undertaken for the most fundamental reason, to prevent the loss of sovereignty over state territory. It stands to reason that actions to reintegrate and regulate this territory would have long-term effects on central state authority. Stein Rokkan describes two types of conflicts that arise as states attempt to centralize (see chapter 1). One poses the central state against regional leaders or social groups who resist centralization. The second conflict emerges as industrialization pits agrarian landowners against rising industrialists.[6] The Civil War involved both types of conflict. And its resolution determined how the central state would manage its territory and resources.

Barrington Moore argues that contention between the centralizing state and large agrarian landholders in the United States was critical in determining whether the state became a democracy or dictatorship.[7] Most farmers in the United States were disinclined to forcefully resist democratic state centralization. They had made the transition to nascent forms of capitalism soon after European settlement, and they were not enmeshed in institutionalized relations with an old regime or church.[8] Slave-owning southern planters, however, were dissatisfied with the national government's actions to promote their region, and they helped push the South into the Civil War. After the South rejoined the Union, however, planters actually encouraged central state building. "When Southern 'Junkers' were no longer slaveholders and had acquired a larger tincture of urban business and when Northern capitalists faced radical rumblings, the classic conservative coalition was possible."[9]

The war made it evident that the government was capable of intervening into the economy intensively and effectively. The war also increased political tolerance for economic regulation and economic aid.[10] After the war, the central state was no longer required to protect slavery to maintain the Union. It could devote itself to economic development policies under the rising conservative coalition, policies that included internal improvements such as flood control.[11]

Civil War Destruction

The Mississippi during the Civil War became a weapon as well as a highway. The Corps of Engineers suspended its work clearing debris. Both armies attempted to disrupt enemy forces by deliberately sinking ruined ships and debris in the river channel and by destroying levee lines. General Ulysses S. Grant's forces destroyed levees and tried to alter the river to bypass Confederate guns at Vicksburg. Battles, Union river patrols, and flooding also disrupted cotton and sugar cropping along the Mississippi and halted all commercial shipping on the river. But neglect of levees produced most of the damage to farmlands, as southern states suspended levee taxes and as slaves fled to Union Army camps or went with their owners to the front lines.[12]

Reconstruction and the River

After the truce, southern flood control activists hoped the federal government would give aid to rebuild war-damaged levees. Republican

President Andrew Johnson and Secretary of War Edwin M. Stanton professed support for federal repairs to levees and to the river channel as part of their mild attempt to reconstruct the South. Members of Congress introduced a number of bills to provide this aid. The Corps of Engineers reported on war damage to the Mississippi's levees, and its crews restarted work on the river channel to improve navigation.[13] But Republican Reconstruction Radicals in the House of Representatives impeached President Johnson in 1868, in part for his moderation toward the South, although the Senate failed to convict him. Disputes about Reconstruction during subsequent years often distracted Congress from internal improvement legislation.[14]

The impoverished postwar South had difficulty competing with the intact economies of the North and the West to attract capital investors. During Radical Reconstruction (1867–75), Republicans whom Congress had installed as governors in southern states attempted to win local support by granting tax breaks to private companies to encourage them to build canals, river projects, and other internal improvements. Few local businesses were able to respond to such offers or to bond sales, and few outside investors found them appealing. Levee districts had always depended on locally managed financing schemes. But most districts were bankrupted, reorganized, and taken over by Reconstruction state governments, and they did little levee work for decades after the war.[15]

Southern development was complicated by concerted competition from northern merchants and by the growing postwar polarization of the Democratic and Republican parties. Northern merchants had a rising consciousness of themselves as capitalists. Merchants in Philadelphia, Boston, and several large midwestern cities tried to gain advantage against each other and against other regions by working to change national policies about southern indebtedness, southern land confiscation, and import tariffs. Some were particularly interested in gaining control over a renewed cotton trade.[16] With party ideologies and party alignments across the sections in flux, most members of Congress therefore voted according to their own local and regional interests on these and other economic development bills. For example, in 1866, the $3.5 million rivers and harbors bill designated only $75,000 for the South.[17]

Conventions Resume, River Appropriations Slowly Increase

In the lower Mississippi Valley, business-minded planters, land investors from the North and from Europe, and their merchant allies in New Orleans and St. Louis lined up against the northern merchants. They promoted federal aid for flood control and navigation to establish southern control over cotton shipping and marketing.[18] Recent generous aid for northern infrastructure, and the migration of northern Republicans to the South after the war, encouraged even many conservative southern whites to demand federal aid.[19]

Some also envisioned levee building as one means of solving the new-found problem of finding farm laborers. Before the war, few landowners or their slaves had lived in flood-prone bottom lands. By encouraging freedmen to settle on leveed bottom lands as tenants or sharecroppers, planters could transfer to the farm workers some of the risk of losing crops to flooding. This strategy also made workers and their families personally vulnerable to floods.[20] John Barry reports evidence that early forms of sharecropping were developed by the very planter families in the Yazoo-Mississippi Delta who organized levee districts and led lobbying efforts for federal flood control aid. Improved levees would aid navigation, allow owners to consolidate the emerging sharecropping and tenancy systems, and expand cropping to new areas in the fertile floodplain.[21]

River improvement conventions and other forms of river advocacy resumed, with planters demanding a new emphasis on flood control for the lower Mississippi River, south of Illinois. St. Louis boosters argued that there would be no more rapid and certain mode for Congress to cancel the public debt than through Mississippi River development.[22] The governor of Louisiana, J. Madison Wells, who had been a Unionist during the war, published a pamphlet supporting aid. It stated that landowners suffering from war losses and the "destruction of slavery" had become unable to maintain levees that protected crops vital to foreign trade and vital to the livelihood of the people whom the northern armies had freed from slavery.[23] The Louisiana Sugar Planters' Association, the Mississippi legislature, and Mississippi levee district boards sent memorials to Congress. They reported that neither planters nor state governments had the funds to maintain levees, and they petitioned Congress to take the river's levee system under federal control.[24] Similarities in the format and rhetoric of the river conventions throughout the valley reflect the circulation of leaders and of ideas.[25]

The proportion of rivers and harbors aid designated for the South increased, ranging from 15 to 26 percent from 1871 to 1878. Southern states were readmitted to Congress in 1868. After extensive flooding in 1874, a delegation from a New Orleans convention that was called in response to the floods lobbied for action. Congress created a commission headed by General Gouverneur Warren to devise a "permanent plan" to reclaim flood-prone lands.[26] Army engineers reporting for the commission, including E. H. Abbot, concluded that uncoordinated levee building by local levee districts on the lower Mississippi River had produced a defective system. They proposed creating a set of regional districts supported by federal aid to build the general system of levees that Humphreys and Abbot had recommended in their 1861 survey. On the basis of this report, Representative Randall L. Gibson of Louisiana managed in 1875 to create a House committee on Mississippi River levees. Northern House members, however, were becoming increasingly opposed to subsidies. A bill with proposals from Warren's levee commission failed, despite support from Republican President Ulysses S. Grant (1869–77).[27]

Still, the flood control cause made incremental progress. Members of Congress from the lower Mississippi River were persistent, and Republicans were not unified in opposition. Negotiations associated with the 1876 presidential election and the Compromise of 1877 may have encouraged reluctant members of Congress to begin flood control aid. Given the lack of full records, the effect of these negotiations on national levee policy is likely to remain in dispute.[28] Regardless, growing southern power in Congress made it increasingly likely that Congress would provide assistance for levee building on the lower Mississippi.

The Compromise of 1877

By 1874, the Republican governors installed in southern states during Radical Reconstruction had become severely undermined. With President Grant providing little leadership, southern moderate and Radical

Republicans were at odds. Democrats won the majority in the House of Representatives in 1874. By 1876, the only Reconstruction administrators remaining were in Florida, Louisiana, and South Carolina.[29] Negotiations that ended Radical Reconstruction came to be called the Compromise of 1877.

In the 1876 presidential contest between Democrat Samuel Tilden and Republican Rutherford B. Hayes, Democratic and Republican election officials in the three states with Reconstruction administrators produced conflicting vote totals for their states. This left the presidential race undecided. Hearing rumors that Republican Party chairs had suppressed Tilden victories in Louisiana and Florida, northern Democrats in Congress and several newspaper editors threatened violence unless Tilden was seated.[30] Congress approved an independent commission of legislators and Supreme Court justices to settle the election.[31] After some party maneuvering, the commission voted for Hayes.

Tilden supporters delayed House proceedings with an eye to blocking certification of the electoral votes. New negotiations opened at the Wormley Hotel in Washington, D.C. Apparently, Republicans agreed to southern home rule, and Democrats agreed to recognize rights for African Americans, among other things. Historian C. Vann Woodward attributes an active role in the negotiations to southern Democratic leaders who pressed for aid to the Texas and Pacific Railway and levee aid for the Mississippi River. Other historians argue that southern whites' interest in regaining home rule was a much greater motivator. In addition, they contend, northern as well as southern Democrats favored ending Reconstruction, and northerners' interest in protecting southern freedmen was weakening.[32]

Flood control aid may or may not have been an important part of the secret Wormley Hotel discussions. But it had been a recurrent theme in North-South politics and had become a priority for many southern leaders. In the next section, I discuss Congress's informal approval of flood control aid soon after the Compromise of 1877.

As president, Hayes tried vainly to build a Whiggish southern Republican Party, signing bills for a variety of southern infrastructure projects.[33] By emphasizing the affronts of Republican Reconstruction, however, southern Democrats drew increasing numbers of whites to their party.[34] In winning the midterm elections in the South in 1878, Democrats consolidated their party's southern power.[35] Conservative southern Democrats found themselves sharing party power with a rising Whiggish-industrial faction and even adopting that faction's positions in favor of federal aid to the South.[36]

Early Federal Aid for Levees in the Interest of Navigation

Aid for levee building came in cautious steps and only after members of Congress were persuaded that levees could help improve navigation. James Eads—an independent engineer in a long-standing rivalry with Chief Engineer Humphreys—had managed by 1876 to deepen the channel of the south pass at the mouth of the Mississippi River by building jetties to direct the flow. In promoting his success to admirers in the public and in Congress, Eads proposed building modified jetties all along the lower river, making cutoffs, and temporarily confining the river with levees to help concentrate the flow and deepen the channel. General Humphreys countered by endorsing the Warren Commission plan, which he represented as proposing to retain the remaining natural outlets and to build a system of permanent levees. Although Eads attacked the Warren Commission plan as a program to create outlets, both plans featured levees as integral to the goal of deepening the navigation channel. Flood control advocates on the House Committee on Levees began to promote this conception of levees to counter northeasterners who opposed any spending on the Mississippi and Republicans from the upper Mississippi Valley who saw no need to ally with the flood control cause. Some of those supporting Eads proposed a bill to create a commission of civilian and military engineers independent of the Corps of Engineers, with Eads presumably as chair. Humphreys took this as an insult to the Corps. Legislators in favor of the commission saw it as an opportunity of decreasing their own reliance on the Corps' advice. Debates again focused on the constitutionality of aid that might benefit private landowners rather than navigation. Representative William A. J. Sparks from Illinois argued, "We cannot and should not draw upon the Treasury to protect 'adjacent alluvial lands,' for such work would be local. The lands are the property of private citizens and within the sole control and under the jurisdiction of the States in which they are located, and there is no warrant of national authority for the expenditure of money for any such purpose."[37]

Repeating the sentiments expressed in activist petitions and at river conventions, some southern members of Congress argued that a plan of river improvement would better unite North and South by reconstructing the damaged South.

Mentions of levees in the bill were largely refocused on their use for navigation, and in 1879, President Hayes created the Mississippi River Commission as an advisory group of civilian and military engineers connected to the Army Corps of Engineers to design plans for the lower Mississippi River. Eads became a commission member, but northern critics saw some of Hayes' early appointments to the commission as evidence that Hayes intended it to promote levees for flood control.[38] Once the commission was formed, states along the lower Mississippi formed additional levee districts and empowered each district's board of commissioners to act as state government representatives to the Mississippi River Commission.[39]

Although the Mississippi River Commission had been empowered to plan for flood control, critics in Congress wrote appropriations bills from 1881 to 1890 allowing levee building mainly for the purpose of improving the navigation channel. Charles Camillo and Matthew Pearcy argue that in fending off critics, commission members came to favor a policy of "restraint in the interest of navigation," building levees that they believed were large enough to constrain ordinary floods to deepen the channel but not large enough to block more destructive floods from alluvial lands. After floods in 1881 and 1882, leaders of levee districts in Arkansas, Louisiana, and Mississippi, including planter W. A. Percy, attended commission meetings to plead for aid. The commission members provided some aid to repair crevasses in the locally built levees. With limited funds arriving from Congress, however, the commission gradually adopted a policy that discontinued the scientific study of outlets and other options mentioned in the Humphreys and Abbot survey but that also failed to satisfy Eads, who quit the commission in 1883.[40]

Despite the commission's caution, critics in Congress would later argue that from the 1880s on, members of river committees in Congress informally directed army engineers to assist levee districts in building levees and in closing large crevasses during floods. According to these critics, members of Congress understood that navigation bills for the Mississippi River in practice delivered some funds for levee work that benefited private landowners. This form of levee aid persisted as an open secret in Congress until the first Flood Control Act in 1917.[41] Republicans could not be seen publicly supporting aid to southern planters. Indeed, the occasional Republican still won election to national office by "waving the bloody shirt," recalling southern Democratic aggressions that led to the Civil War. On the other hand, Democrats could not be seen advocating the intervention of the state into the economy through tariffs, subsidies, or rivers and harbors improvements, positions that were still associated with the Republican Party. Landowners along the Mississippi River also found advantage in avoiding political scrutiny by requesting small projects directly from the Corps-affiliated Mississippi River Commission rather than from Congress.[42] This indirect approach to policy making therefore reinforced the localism of the Corps' river management work.

Conclusion

The Civil War extended the central state by changing expectations about how the government should intervene into resource development and by changing relationships between the central state and the South. River activists set flood control on the agenda by pressing for advantage after the contested presidential election of 1876. In the 1880s, Congress decided to provide levee aid for the lower Mississippi River. The flood control program and other land programs increased federal responsibility for economic growth. These programs also allowed local elites considerable influence over the way government work would be completed. Skowronek shows that civil service reforms and other measures to increase professionalism managed to remove regional and party control over many of the federal government's activities in the late nineteenth century.[43] But localism survived in federal land development programs like the flood control program.

The Civil War and the subsequent failure of Reconstruction established new accommodating relationships between the central government and the South, and it changed relationships between the two major political parties. Southerners lacked power in Congress after the war, but they managed on occasion to unite around specific demands, including flood control.[44] Flood control aid was a small step in the Democratic Party's shift toward promoting government economic intervention. More importantly, it affirmed for southern Democratic leaders the strategic value of unifying their delegation behind narrow demands and sustaining those demands over the years.

People had become accustomed to government taking on extraordinary powers during the Civil War. Even so, the decentralized organization of the postwar flood control program was considerably less directive than the government's wartime economic activities had been, and it became typical of the federal government's cautious economic interventions during the coming decades. In response to state governments' practice of impeding interstate river trade, the 1824 Supreme Court ruling in *Gibbons* had defined federal responsibilities to provide for interstate commerce. Congress interpreted this as a requirement to improve navigation channels. In response to the delicate compromise struck between North and South after the Civil War, congressional decisions in the 1880s informally broadened federal powers to aid economic production.

chapter 6

The Sacramento River

MINERS VERSUS FARMERS

> Since the discovery of gold, the "washings" render [the Sacramento River] as muddy and turbid as is the Ohio at spring flood—in fact it is perfectly "riley."—William Henry Brewer, *Up and Down California in 1860–1864: The Journal of William H. Brewer*, 294–95

Access to Gold

The integration of the far western territories and reintegration of the South after the Civil War produced a new phase of regional competition. Events that had added the far western territories—the Mexican War and the Gold Rush—were not as critical as the Civil War in defining central government powers, but admitting the western territories as free-labor states changed national politics by reducing the political influence of southern slave states and increasing southern worries about regional economic competition. Despite this competition, elites in the Mississippi and Sacramento valleys struck a similar note of grievance against the federal government for its failure to provide them adequate resource development aid and they joined together in promoting flood control aid.

As the site of the major strikes of the California Gold Rush, the Sacramento Valley was the earliest area of the Far West to be developed intensively. This chapter describes the haphazard beginnings of U.S. rule there. After the initial rush of individual gold miners to the Sacramento area in 1849, mining became a capital-intensive industry. Miners developed nozzles to blast away hillside deposits and redirected mountain streams to power these operations. Catastrophic debris flows onto valley farms eventually set farmers against miners. Unlike the lower Mississippi Valley, where elites were culturally and economically united in their commitment to agriculture, elites in northern California came into conflict when mining activities began to harm agriculture. Chapter 7 describes the resolution of the conflict between miners and agriculture in the courts and the beginning of concerted lobbying for federal aid to redress the damages from mining.

Returning to Stein Rokkan's characterization of state centralization, the integration of the Mississippi and Sacramento river territories provoked conflicts between elites in peripheral regions and the central government as well as conflicts between agricultural landowners and industrialists.[1] In the lower Mississippi River valley, agrarian landowners' interests dominated. The competitive pressure they felt from the emerging industrial economy was largely experienced as pressure from outside their region. By contrast, California's economic base was established much later, when industrialization had advanced. Entrepreneurs quickly committed themselves to intensive "industrial" techniques, even in farming. In California, conflicts over the effects of industrialization arose between two industrializing sectors, mining and farming. Industrialization also pitted small farmers against large farmers (especially swamplanders) and small miners against large mining and water companies. Once it became clear that the future belonged to intensive agriculture, the Sacramento Valley produced a flood control activist movement similar to the movement in the Mississippi Valley.

The Golden Sacramento River: The Physical Setting

The Sacramento River and its tributaries revealed the veins of gold that first attracted the forty-niners to California. These rivers also created fertile delta areas that eventually made the valley one of the country's most productive fruit and vegetable growing regions. This placed farms in the path of mining debris. The Sacramento River drains the country's deepest snowpack and is subject to the largest fast-rising floods of any U.S. river system. The river descends into the northern Sacramento Valley from the Cascade Mountain Range, meeting with the gold-bearing tributaries of the Feather and American rivers above the capital city of Sacramento. Below Sacramento, the river enters the Sacramento-San Joaquin Delta, where the Sacramento and San Joaquin valleys meet.

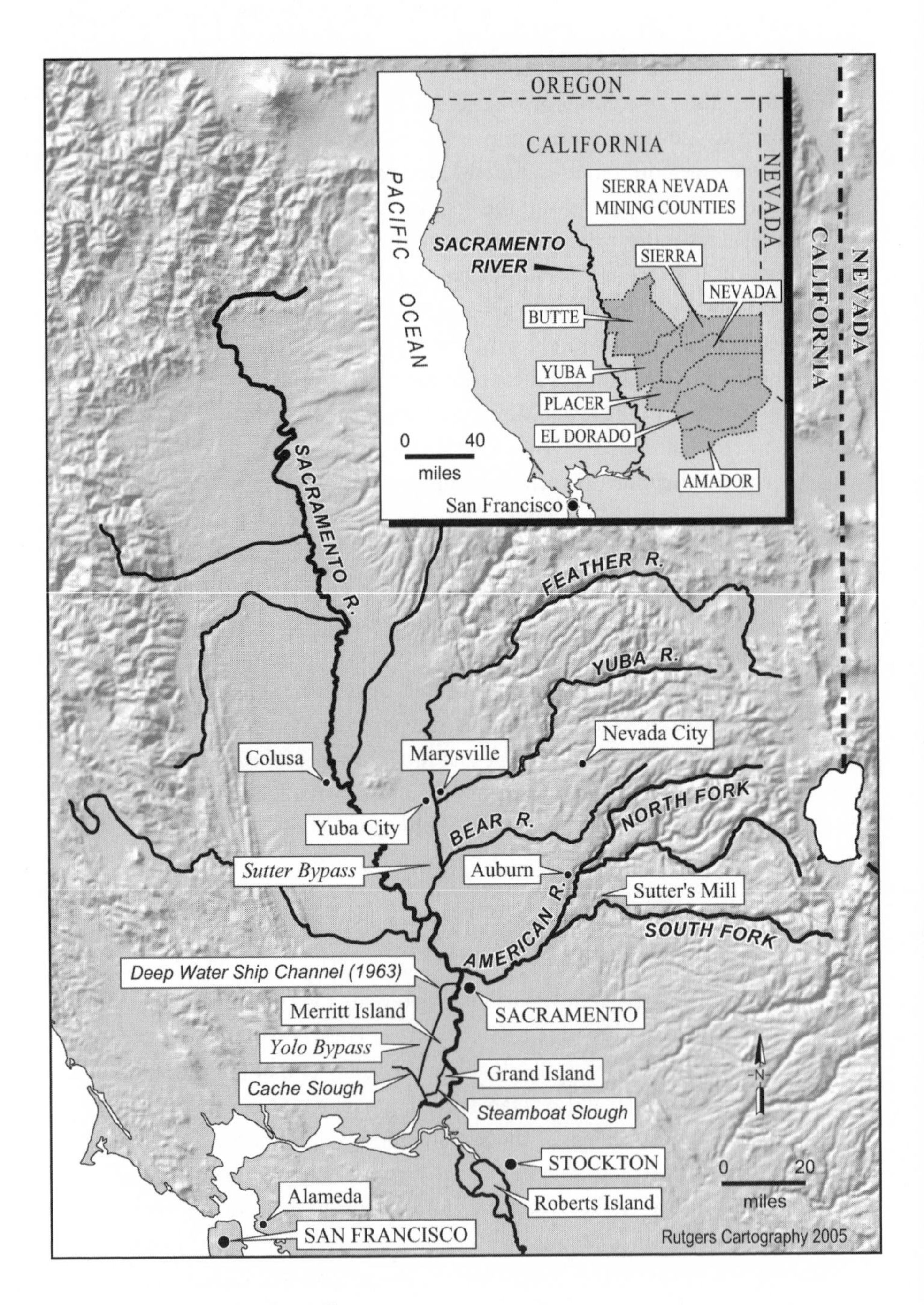

MAP 2. Sacramento Valley

Together, these two valleys form the Great Central Valley. Delta waters exit to the sea through a series of bays leading out past San Francisco.[2]

Before Europeans arrived, the Sacramento River had a clear, if not deep channel. The valley had wooded marshlands and open wetlands. These areas would flood during winter rains and spring snow melts, often turning into an "inland sea."[3] Indigenous peoples and European trappers and traders visited the lowest areas of the valley only during the six months from summer to fall, when the valley was not flooded from winter rains. For safety and comfort, they camped on the natural levees of the delta waterways. Spanish and Mexican settlers in northern California seldom ventured that far inland.[4] Gold Rush settlement occurred so quickly that few settlers or government officials therefore learned about the valley's dramatic seasonal floods.

Conquest by War and by Gold Rush

The United States faced much weaker opposition from foreign nations and indigenous groups in the Far West than it had in the lower Mississippi Valley. As a result, infrastructure building became politically important only after conquest, in the context of economic, rather than military, disputes. Spanish and Mexican colonists had claimed only a few vaguely defined ranchos in the Great Central Valley. They considered the area strategically unimportant, and very few settlers had moved there. Some lived in the coastal Spanish missions founded to convert the indigenous populations, but most lived on southern California cattle ranchos.[5]

Mexico rejected several offers by the United States to buy its territories in the Southwest and Far West. The United States handily won those lands through war. For the United States, the timing of the Mexican War was fortuitous. In September 1847, U.S. troops took Mexico City and ended the war.[6] On 24 January 1848, the foreman at Sutter's Mill on the first American settlement in the Sacramento Valley discovered gold traces in a millrace on the American River.[7] Nine days after the gold discovery, but before word was out, representatives of the United States signed the Treaty of Guadalupe Hidalgo with Mexico. This added lands to the United States that would become the states of California, Nevada, Utah, and New Mexico.[8] The gold rush began in

1849. Settlers created a constitution in September 1849 and elected a governor in early 1850. The U.S. government formally admitted California as a state later that year.[9]

Gold was compelling because the United States and other countries based their currencies on gold reserves. By the late nineteenth century, the United States produced nearly 45 percent of the world's total extraction of gold, with California leading production.[10] If gold had not been discovered, it is likely that white settlement of the swampy Sacramento Valley would have been delayed for years.[11]

Putting Resources to Productive Use under Nominal U.S. Rule

United States authority was barely felt in the gold region. The understaffed office of the United States military governor prosecuted few gold seekers for failing to file formal claims on unsurveyed U.S. government lands. Local practices that developed in the stead of government authority privileged the mining claimant's right to develop a parcel, with little regard to the effects that mining might have downstream. Miners founded haphazard settlements in the mountains and regulated their own gold claims. Working from traditions of Wales, the Harz Mountains, northern Spain, and elsewhere that assured miners' autonomy, California miners created "unconscious socialism."[12]

Yeoman farmers were rare, and independent miners flourished only briefly in northern California. Investors in mines and farms managed to consolidate large land holdings soon after the beginning of the Gold Rush. As well, nearly all of the California parcels that people had claimed as colonial land grants from Spain and Mexico were large, some as large as counties, although the long-term effects of colonial grants on land tenure were indirect.[13] Many large colonial claims along the Sacramento River and its tributaries were recognized by the United States.[14] By the 1860s, ownership of most colonial grants had passed from colonial ranchers to rising entrepreneurs in mining and farming. Those new owners were more adept at shaping and responding to changing legal and market conditions and more interested in pursuing intensive production. Colonial land grants that were broken up were often consolidated into new largeholdings that included grants from the U.S. federal and state governments.[15] The result was higher rural land concentration in rural California than was typical in other rural areas of the United States.

Unsuccessful gold miners and other migrants meanwhile settled the Sacramento and San Joaquin valleys below the mountains to grow and ship produce, sell provisions to the miners, and ship out gold. European and American landowners in the northern valleys found a ready agricultural workforce of Chinese and Mexicans and other "foreigners," whom whites had forced from the best gold diggings.[16] Sacramento became the major river port city for the gold trade, the seaport city of San Francisco expanded, and it became clear that the real money was in farming and provisioning, not mining. By 1850, river traffic in the widest portions of the Sacramento River included large steamboats. In 1854, the California Steam Navigation Company began to force out, or buy out, competitors.[17] Navigation was difficult, but not impossible during these early years.

Swamplanders

California entered the Union in 1850, the year that Congress approved swampland grants to all public land states then in the Union (as discussed in chapter 4). The swampland program enhanced land concentration near rivers throughout the state, and it produced the most important transfer of land in the Sacramento–San Joaquin Delta.[18] In 1862, Gov. John Downey in his annual message complained that there was no evidence of "practical benefit" resulting from the swampland grants. He blamed swamplanders for being unwilling to pay their portions of levee building costs.[19] Many swamplanders had indeed bought parcels for speculation rather than productive use, but haphazard and often corrupt administration of the program had also confused land titles. It took decades for swamplanders to undertake reclamation and economic development of the forbidding parcels that they claimed in the San Joaquin Valley and the Sacramento Valley.[20]

Mining and Floods

Many of the new settlers viewed floods in the 1850s first as unusual events, then as conditions to be accommodated, and finally as threats wrought by reckless upstream miners. In the winter of 1850, floods threatened the low-lying city of Sacramento, which was then competing for the mining trade against small settlements on higher ground along the American River.[21] Migrants from the Mississippi Valley familiar with flooding evidently encouraged some city residents to build private levees.[22] A citizen's committee designed a river survey and made plans for a larger city levee. A joint citizen-city government levee commission completed the survey, and voters approved $250,000 in taxes to build the levee. Over the next few years, the cities of Sacramento and Stockton tried various schemes to accommodate to life in the swamp. They built elevated turnpikes, reclaimed nearby swampland, improved street drainage, raised street levels, and planked the streets.[23] Changes in the ownership of water, the ownership of mining claims, and mining techniques would deliver a new set of problems to the valley.

Access to water for mining operations was commodified soon after the Gold Rush. This step began the transition in mining to concentrated mine ownership, intensive technology, and transformation of the rivers flowing to the Sacramento Valley. In the early years, when mid-level mountain streams dried in the spring, miners hauled their diggings downhill to the rivers to sluice out the gold. It was a cumbersome process, but the miners had free access to water. Taking advantage of the need for water, a group of enterprising miners formed the first water combines with Nevada City merchants, in 1850. They channeled the waters of higher, year-round streams down to mining sites and began selling the water to other miners.[24]

Meanwhile, placer mining techniques that could be managed by individual miners were becoming ineffective. Within two years of the 1849 Gold Rush, the low-cost and small-scale technology of gold pans, one-man sluices, picks, and shovels had captured most of the free-flowing gold in Sierra Nevada rivers.[25] Successful claimants and newly attracted outside investors consolidated mining claims and formed larger mining operations. White American and European miners employed by these operations resisted peonage. They also steadily harassed the indigenous peoples, African American slaves, Pacific Islanders, Mexican peons, and, in later years, Chinese indentured servants, whom mine owners brought in as laborers. This white resistance kept wage costs very high in the northern Sierras.[26] Other miners seeking to sustain their independence began working in narrow ravines. There, they found gold streaks along buried beds of gravel, which miners called "dead rivers." Geologists eventually traced these vast, Tertiary period waterways across large portions of the Sierras.[27] To gather this sparsely distributed gold, and to reduce labor costs, miners adopted increasingly invasive and capital-intensive methods.

Miners who ran ditch operations soon refined hydraulic mining techniques. These operations passed water under pressure through hoses with nozzles to blast away hillsides.[28] The first nozzles allowed miners to wash away the top layers of earth, moving 50 to 100 cubic yards of earth each day. Mine owners could employ a few miners where hundreds had worked before.[29] Hydraulic mining operations began to deposit debris in mountain canyons. The heaviest deposits were in the Yuba, Bear, and American rivers just below the productive mines of Sierra, Nevada, Placer, and El Dorado counties.[30] Valley dwellers knew about this debris, which had visibly reduced the Yuba River's flow by 1858.[31] As water costs rose and returns declined, miners held mass meetings and occasionally destroyed the ditches of water enterprises to protest their fees. Yields from hydraulic mining and other mining methods declined in the late 1850s, and new metals strikes in Nevada and Idaho lured away Sierra miners.[32]

Finally, during floods in the winter of 1861–62, debris piles burst, sending giant flows of clay and rock onto valley farms along several rivers.[33] Nearly thirty-three inches of rain fell in two-and-a-half months that winter, carrying debris down with it. According to one settler, the new lake formed in the lower valley was sixty miles wide. Steamboats moving cattle from flooded areas to the hills were able to paddle as far as fourteen miles from the riverbed. Most of the city of Sacramento was under water for three months. City levees that had remained intact trapped floodwaters that had broken through east-side levees. The state legislature relocated temporarily from Sacramento to San Francisco.[34] These events disturbed the ready assumption that mining benefited all California residents.

Mining was the main business in mountain towns. Mountain residents had treated mining as more important than farming, and they supported the rights of miners to use resources as they wished. Miners claimed all the potential farmlands in the mountains. Hydraulic miners sluiced away local roads and pastures, and pit miners dug tunnels that literally undermined towns.[35] Many mountain residents therefore sympathized with the valley farmers but expected them to bear with the side effects of the state's dominant industry.

Valley residents had mixed sentiments about the damage they suffered from hydraulic mining. Some said that miners were imposing risk on valley residents by enhancing the natural floods. They began to argue for various forms of recourse. Others said that valley farm and city residents should clear the debris themselves, as their livelihoods depended on the gold mining trade. The editor of the chief mining newspaper later commented that when miners were "the best customers of those who cultivated the soil, buying at liberal prices and consuming largely of their products, the farmers raised no objection to the evil they now complain of."[36] Splits among valley residents delayed formation of a protest movement. The split between mining capitalists and valley critics delayed activism for federal aid that might have benefited both industries.

Federal Navigation Improvements on the Sacramento

An army engineer for military operations had been posted to California even before it became a state. This officer—who later also acted as secretary of state for California—first ordered engineering studies for regional defense in 1846. Army engineers built fortifications near San Francisco in the 1850s, and they began navigation surveys of the Sacramento valley in 1855. In the absence of credible foreign military threats, Corps engineers' work soon centered on clearing navigation channels in the Sacramento Valley, a task with purely commercial ends.[37] By 1866, debris had ended the infamous side-by-side steamboat races along the Sacramento River. Large steamboats only occasionally made it upstream to the port of Sacramento.[38] Congress provided appropriations to improve river depth and reduce sand bars in 1875, and the Corps completed its surveys and did snagging and dredging work. But navigation problems persisted in the valley because hydraulic mining operations continued.[39]

Origins of an Agricultural Giant

During the early years of U.S. rule, farmers in the Sacramento Valley primarily supplied the gold miners and generally viewed their interests as shared. The first farms to produce surpluses for outside markets were north valley barley farms. Vast, mechanized "bonanza wheat farms" in the lower Sacramento Valley, held largely by absentee owners, peaked from the 1850s to the 1870s.[40] As early as the 1870s, some farmers on the lowlands just west of the Sierras were growing and shipping the high-value row and orchard crops that characterize valley agriculture today.[41]

California state land grant programs fostered land concentration. The Glenn, Bidwell, Gerhke, and Cone farms in Sacramento Valley and the Moulton farms in the Sacramento-San Joaquin Delta ranged from 18,000 to 100,000 acres, mostly in wheat. Henry George and other reformers attacked wheat farming in California for encouraging land concentration.[42]

As a consequence of this pattern of land tenure, arable and overflowed lands outside of the city of Sacramento were sparsely settled. Farmers there built few levees in the years before Swamp Land Act titles were settled. One settler on Grand Island in the delta built levees as early as 1851. When Swamp Land Act sales were authorized by state law in 1855, he and landowners of neighboring islands built small levee systems on their own or in cooperative efforts.[43] Privately built levees on Grand, Merritt, and Roberts islands were sufficient to allow profitable farming during some years.[44] Owners in most areas refrained from draining wetlands, however, because the state government's title to swamplands remained too confused to justify costly land reclamation.[45]

Capitalists Take over Mining and Form a Political Bloc

In the 1850s and 1860s, most miners adopted capital-intensive mining techniques such as hard-rock quartz mining, river dredging, and hydraulic mining, making the mining economy dependent on capitalized joint enterprises.[46] Booms and depressions alternated, depending on technological developments, fires, and droughts. A drought from 1862 to 1864 stopped hydraulic mining and slowed the buildup of debris in mountain canyons.[47] Operators had begun to consolidate small water claims as early as the 1850s, and the drought hastened this consolidation.

The end of the drought in late 1864, the return of many miners from the Nevada Comstock silver rush in 1865, and improved mining techniques renewed interest in mining the Sierras. Investors from San Francisco, eastern states, and England financed a second phase of hydraulic mining. New nozzles blasted the compact lower gravels, and gold again arrived in quantities into San Francisco.[48]

This phase broadened the gap between "kid glove" mine owners and mine workers, and it strengthened the political pull of miners in state politics. Between 1860 and 1868, the number of operating mining companies dropped by one-third. Owners became increasingly wealthy, increasingly powerful in state politics, and increasingly distant from the mines. Mine workers had little influence over merchant-dominated town politics or over state politics because they usually stayed in mountain towns only briefly, whether successful or not.[49] San Francisco became the port for gold shipments and supplies and the state's center for banking and finance. Its merchants and absentee mine owners acted as the core of a pro-mining voting bloc in state politics. Mining remained the state's dominant export industry and top employer as late as 1863. Even after mining declined, it was years before farmers' complaints about debris damage became important in the regional and state political realms that were dominated by San Francisco.[50]

The second stage of hydraulic mining in the 1860s brought unquestionably serious damage to the valley. Debris flowed down regularly from mountain canyons. By 1868, the beds of the Feather and Yuba rivers were higher than the streets of Marysville, a trading city at the confluence of those rivers. Marysville residents organized a public-private effort to build levees around the downtown area. Farmers above Marysville built a seventeen-mile-long levee, which floods broke.[51] On flatlands that the Yuba and Bear rivers had covered with rocks and mud, some farmers built levees, channels, and drains and opened closed drainage sloughs. Others simply moved.[52] Farmers and townspeople around Marysville and Yuba City who lived in the heaviest debris areas in the state were also those most dependent on trade from the mines, and they resisted taking political action against the mines.

Conclusion

The extension of U.S. rule to the Far West provoked renewed competition among the regions, as western settlers and outside investors and traders attempted to take advantage of California's mineral and agricultural resources. With an easy military victory over Mexico, the U.S. government directed its military officials to administer civil rule. The speed of settlement during the Gold Rush overwhelmed the provisional federal authorities in California. Northern California capitalists were able to quickly build wealth from mining, agriculture, and rail development. This economic strength, plus the state's growing power in Congress and its physical distance from the settled East, made it unlikely that the federal government would impose decisions about resource development.

Before the 1870s, valley farmers and other residents did not organize to demand federal government aid for rivers against the damage they suffered from mining. Many farmers and townspeople were politically skeptical of government action, and many still depended heavily on trade with the miners. Only as it became clear that agriculture and water, rather than mining, would become the dominant economic concerns of Sacramento Valley did valley residents begin to seek legal and political solutions to mining damage.

chapter 7

The Sacramento River

CAPITALISTS UNIFY FOR DEVELOPMENT

Conflicts between Mining and Farming become Political

By the 1870s, farmers and most merchants in the Sacramento Valley viewed agriculture, rather than gold mining, as the key to future valley development. They followed the example of activists in the lower Mississippi River by arguing that the federal government had a duty to improve the lands that it had encouraged settlers to use. As in the Mississippi Valley, upstream farmers joined with traders and other business leaders in downstream port cities in pressing both the state and federal governments for navigation and flood control aid. In the Sacramento Valley, however, political activism for navigation and flood control aid began in earnest only after a federal court ended the immediate conflict between miners and farmers in 1884, by restricting harmful mining operations.

This conflict within the capitalist class indicates the key role that environmental control can play in determining how industrialization and class formation proceeds in a specific region. The political mobilization of miners and farmers had effects on the emerging capitalist economy well beyond creating physical infrastructure systems. Hydraulic mining operators were able to reorganize as water monopolies that would have a lasting influence in California politics, and large landowners came to dominate agricultural trade and politics. Although the rise of intensive agriculture in California may seem inevitable, the events described here suggest that it was not and that specific employment practices and patterns of settlement and cropping depended on many political contingencies.

Despite the many differences between the Mississippi and Sacramento valleys, elites in these valleys organized in similar ways to take advantage of earlier federal government commitments to land development. By the 1880s, activists from the two valleys were cooperating in their campaigns for federal aid, and legislators were regularly trading votes for development aid bills benefiting the South and the Far West.

An Attempt at Centralized Planning

The California Legislature attempted to coordinate building of levee systems in the 1860s, creating a Swamp Land Commission and appropriating $200,000 for a plan.[1] Whig-Republicans in California had criticized the poor administration of the federal swampland grant program, but they favored land development aid. They now spoke of the benefits of encouraging large landowners to build levee systems for the Sacramento Valley. Republicans from the city of Sacramento allied with representatives of the neighboring mining county of El Dorado and led passage of this legislation.[2]

This legislation granted California drainage districts powers similar to those of the Mississippi River levee districts, including the powers to tax, to regulate fellow landowners, and to build levees. Engineers appointed by the state were to plan levees for each district.[3] In the first year, landowners proposed twenty-eight districts scattered in the Sacramento and San Joaquin valleys and around San Francisco Bay.[4] The program produced few lasting results. Floods damaged levees, Civil War preparation slowed building efforts, land sales were slow, the fund ran out, and the federal government land commissioner rejected state surveys. Landowners complained about assessments, and districts repeatedly redrew district boundaries. Some landowners refused to pay levee fees, and others were unable to pay. Even if all farmers had paid, revenues would not have been sufficient. The Sacramento city levee fund received a grant and reclaimed a small amount of land, but there was little other evidence of the program's work.[5]

The legislature expanded levee districts' powers in 1864, allowing districts to issue warrants backed by all district assets, including fund holdings, accounts receivable, and unsold lands. This act stimulated more concerted efforts to build levees. By 1865, the state had approved fifty-four districts, fourteen districts had received funds for reclamation, and surveys were continuing in thirteen other districts.[6]

Radical Localism Becomes State Policy

Partisan politics and opposition from landowners and state officials ended the state's experiment in centralized planning in 1866.[7] Democrats on the Swamp Land Committee of the State Assembly attacked the Swamp Land Board and argued for local control. Some politicians attacked the use of deficit financing. Some districts did accumulate large debts, which the state had to pay from its fund from the federal swampland grant program.[8]

Poor reclamation results in the American River basin near the city of Sacramento in particular caused legislators to form a low opinion of the board, according to the *Sacramento Daily Union*. Sacramento residents had paid perhaps as much as $1.5 million for their own levee building projects before 1861. They resented a proposal by the Swamp Land Commission to tax the city to pay for failures of the larger drainage district surrounding the city.[9] Republicans were able to protect the board plan by passing legislation placing state swamplands and money from the prior sale of swamps into county trusts for reclamation.[10]

With the wheat boom in full swing and federal-state swampland disputes being settled, a new measure passed by Democrats stimulated a new swampland rush. The 1868 Green Act removed limits on the number of swampland acres that could be claimed under the federal swampland program in California. It also devolved responsibility for creating districts onto landowners, in their roles as levee district trustees.[11] Nearly all of the state's remaining millions of acres of swamp- and overflow land passed to private owners between 1868 to 1871.[12]

Only one-sixth of the granted swampland was reclaimed by private owners by 1878. Landowners in the Sacramento–San Joaquin Delta built extensive levee systems in the 1860s and 1870s, although most of these failed repeatedly. Farmers often used white contractors, who typically directed a Chinese foreman on subcontract to hire and coordinate a work team of Chinese and Hawaiians. Whites dominated the skilled jobs on levee projects. A number of swamplanders in the Sacramento Valley did eventually build levees and drain swamps, using the land grants to establish intensive agriculture in the bottomland areas.[13]

Farmers Disagree over Strategy

Calls for government action arose from farmers in several areas of the Sacramento Valley, but farmers divided their efforts between the political arena and the courts. There was early but brief legal action in Butte County, prolonged legal and political activism centered in Marysville and Yuba City, legal and political activism by local landowners near Colusa, and political activism in Sacramento.

The first damage suit by a farmer against a hydraulic mining company was sponsored by farmers along Dry Creek in north Butte County. They had met in 1872 to protest farm damages. Even though the trial attracted wide attention, by choosing a legal strategy centered on damages, these farmer protests failed to create a sustained political movement. The jury of miners and farmers found that fifty companies had worked the area before the present company had and that other miners would continue the work even if this company folded. They also concluded that mining predated farming in the area and that the farm in question was worth much less than the mine was worth. When farmers threatened another suit during the 1874 floods, the mine's owners ended local protests by negotiating to buy damaged farmlands and build a debris ditch. In the coming years, small landowners elsewhere would repeatedly file suits to ban hydraulic mining.[14]

The environs of Colusa, Marysville, and Yuba City suffered the greatest harm from mining. Most small landowners and many townspeople in these areas distrusted the swamplanders who advocated government flood control aid. They also suspected that government works would only encourage miners to continue dumping. Farmers near the Yuba and Feather rivers met in 1874 to form a movement. Residents did not meet again until the next winter, when debris overtopped Marysville levees and filled basements and ground floors with debris. Encouraged by the meetings in Marysville, a Democratic assemblyman from the area, D. A. Ostrom, introduced bills to end hydraulic mining. Mining supporters in the legislature easily defeated these bills as well as requests in 1876 for investigations. A group of farmers in Yuba City pooled their money in 1876 and filed for an injunction in the name of James Keyes against the Little York Gold and Water Company and eighteen other mining, water, and ditch companies and individuals operating near the Bear River. Miners and farmers alike recognized the case as important.[15]

Miners and Farmers Organize as Opponents in Court

These early lawsuits prompted farmers and miners to form permanent organizations, which became the bases for later political activism. In September 1876, a group of miners and mine investors met in San Francisco and formed the Hydraulic Miners Association. They proposed to fight legislative attempts to regulate or ban hydraulic mining and promised to help pay the legal costs of mine owners sued by valley residents. Membership in the Hydraulic Miners Association was open only to owners of mines or ditch companies. Each owner received one vote for every $5,000 of assets.[16] These miners requested that the California Legislature appoint a commission to study the damage done by debris. The legislature's mining committee merely recommended sending a memorial to Congress to create such a commission.[17]

Faced with rising costs and slow progress in the *Keyes* suit, residents from the Yuba City and Marysville area in Sutter County formed the Anti-Debris Association of Sacramento Valley in August 1878. Keyes himself called the meeting to order. C. P. Berry, a member of Congress from Sutter County, argued that farmers must unite to match the statewide strength of the Hydraulic Miners Association. Anti-Debris Association members talked hopefully of gaining a definitive injunction in one court case to ensure success in future suits and bring an end to hydraulic mining.[18]

At the request of the Anti-Debris Association of Sacramento Valley, San Francisco business leaders formed a committee in 1881 to investigate the tributaries and valleys damaged by debris. At Marysville, they found debris had raised the riverbed to the level of the city streets. On the south side of the Bear River, railroad tracks had been raised three times to escape the debris, for a total of eighteen feet. The Keyes and Brewer ranches, which had been under full cultivation, had become "sandy deserts."[19]

The committee concluded that it would be feasible in some river areas to route debris into tule swamps. They also held that protecting Marysville and Yuba City—which lay at the base of a canyon—would require debris basins in the mountains upstream. They recommended that the Anti-Debris Association of Sacramento Valley meet with miners to coordinate political activities of the farming and mining industries.[20] That meeting, on 17 November 1881, proposed little that was new.[21]

Farmers and miners alike traced their rights to federal land grant policies, to federal laws, and to the U.S. Constitution. Farmers argued that Mexican law separated mineral rights from agricultural rights, that Mexican land grants in the valley were agricultural grants, and that mine dumping therefore violated the established right to farm. The mayor of Marysville argued that the federal government had guaranteed the right of settlers to possess agricultural lands and that the government should prohibit miners from disposing waste onto farmland. But state courts had decided early in the state's history that agricultural plots on U.S. public lands could be worked for gold, unless the lands had been patented as farmlands.[22] Hydraulic miners further argued that they had vested or prescriptive rights to dumping in the rivers, which the state and federal governments had recognized by selling miners mountain land for mining.[23]

The public statements of farmers and miners alike were deeply colored by resentment. The farmers told of damaged orchards, falling land values, and the impossibility of planting row crops. Miners argued that cessation of gold mining would "greatly embarrass the General Government" by lessening gold supply. "Even under these conditions, these farmers, although there was still plenty of public land elsewhere upon which to settle, came and planted themselves right in the path of this outflowing sediment, being at the time well advised as to the existence of these usages." The debris was "mere silt" by the time it reached most of the valley. Had "these thriftless husbandmen" distributed the debris over their lands rather than allow it to pile up, they would have increased the land's fertility.[24] According to miners, farmers had exaggerated the extent of damages in their accounts to government and the press. Moreover, the farmers' proposals to have the miners pay for all damages were not much of a compromise.[25] Such declarations resurfaced during later campaigns for federal intervention.

Under Attack, Capitalists Find Common Ground

When for a time class and party politics in California shifted, mining and farming capitalists found reason to work together against small farmers and workers. In 1877, during the depression caused by the panic of 1873, white workers led vicious anti-Chinese riots in San Francisco. They then formed the Workingman's Party, which sought to regulate monopoly capitalists in mining, agriculture, and other industries and to restrict employers' use of Chinese laborers.[26]

As directed by a voter initiative passed in response to the depression, the state legislature convened a constitutional convention in 1878.[27] The Granger organization, with its membership of small farmers, argued that the tax structure benefited large-scale mining and stock raising but discouraged agriculture. They also sought to clear up water rights between the federal and state governments.[28] Workers joined with small farmers to propose various ideas for redistributing the state's water.[29] To head off radicalism, Democratic and Republican leaders united against the Workingman's Party and got a majority of nonpartisan delegates elected to the convention.[30]

The resulting constitution regulated monopolies and restricted Chinese labor, but it failed to reform the water system. Several convention delegates accused the Spring Valley Water Company, which contracted to San Francisco, of conspiring with hydraulic mining and ditch companies to prevent discussion of watershed deforestation and debris damage caused by mining and water monopolies.[31] Historian Donald Pisani argues that even without such a conspiracy, reform was unlikely because so much of the state's economy depended on the existing haphazard system.[32]

An Attempt at Centralized Regulation

Farmers and miners struck a tentative alliance in 1878, when the Sacramento River flooded the entire valley, leaving only Sacramento and Marysville dry. At a special meeting of a State Assembly committee called in Marysville, swamplander William H. Parks suggested that forcing river waters into one channel would scour debris in the channel, protect valley lands, and make it unnecessary to end hydraulic mining. Most small farmers opposed the plan's involvement of state government. Some area miners and farmers at this and later public meetings decided to support Parks's idea.[33]

Pressed by activist farmers seeking irrigation and flood control aid, legislators created a state engineer's office. The new engineer, William Hamilton Hall, made the first authoritative surveys of the river system from 1878 to 1880 in cooperation with local drainage districts and with Corps of Engineers officers.[34] While politicking about water continued, a state Supreme Court decision appeared to reduce the pressure on miners. A judge ruled in November 1879 that the parties in the *Keyes* suit were misjoined because one or more of the miner defendants might be innocent.[35]

Swamplanders meanwhile proposed that the legislature create a drainage district in the capital city area and offered to pay for a Yolo Bypass built through the Montezuma Hills. State engineers surveyed the proposed route and rejected the proposal as too costly and "impracticable" because the sloped area would clog with silt. The engineers did, however, stress the need for a central authority to plan river improvements. In a message to the legislature, Gov. George Perkins called for reclamation plans that would settle the disputes of miners and farmers.[36]

Farmers introduced a bill in 1880 to create a state board to improve rivers and to supervise the reclamation of swamps on private lands. Under a Republican majority, a coalition of legislators supporting San Francisco mine owners, mountain miners, and valley farmers passed the 1880 Drainage Act after weeks of bitter debates. Drainage District Number 1 began work under the presidency of swamplander William H. Parks. The state engineer began a water management plan. By the beginning of 1881, the State Board of Drainage Commissioners reported that brush dams on the Yuba and Bear rivers and eighteen levee projects had been completed for $400,000.[37]

Southern Californians and other opponents of river aid pushed a repeal bill in the Senate in 1881. Southern Californians opposed being taxed to benefit the more populous and wealthy north. The repeal bill created such infighting that the legislature failed to pass an appropriation bill that session. Aid opponents also won in the state Supreme Court, when it ruled that the legislature did not have the power to tax the many to benefit the few. The court declared that debris storage was a private matter separate from drainage improvement. The Drainage Act was therefore unconstitutional on the grounds that it contained more than one subject. The legislative coalition for river development was not sufficiently strong to survive the opposition.[38] Historian Robert Kelley blames the failure of the drainage program on the continued split between Sacramento Valley farmers who sought flood control aid and farmers who sought to end mining.[39]

Federal Navigation Improvements

Over the years, army engineers in California became closely associated with landowners and politicians involved in Sacramento River flood control.[40] The Corps continued to make surveys and do navigation improvement work, as appropriations allowed. But mines continued to pour debris into the rivers, and Corps efforts were insufficient to clear the channel.[41]

Memorials from the California legislature and reports from Corps officers prompted Congress in 1880 to approve planning to prevent mining debris from harming navigable rivers.[42] The head of the Corps' California operations, George H. Mendell, presented the most extensive survey of the mining region thus far in 1882 and requested $500,000 from Congress to fund a detailed plan. Congress directed the Corps to design a mountain dam system to prevent further damage to navigation and appropriated $250,000 to build three dams.[43]

Many valley farmers, however, still sought to ban hydraulic mining. They opposed building large mountain dams, fearing that dams would satisfy government officials, persuade judges to lift injunctions on mining, but fail to prevent further damage. Some sent petitions of protest to the secretary of war. A staff member also reported to the secretary that farmer opposition was high. The secretary concluded that it would be unwise to start dam building until hydraulic mining was stopped, and he ordered further study.[44]

A Decisive Court Ruling

In the end, court action changed the nature of the conflict. Miners and farmers had followed the course of each major lawsuit with interest. They understood that if a judge accepted a particular argument in one case that farmers might be able to successfully sue other mines in other locations using that same argument. The decisive case had implications that went well beyond setting a basis for lawsuits.

On 12 June 1882, Judge Jackson Temple of the California Supreme Court approved a perpetual injunction in one case, prohibiting the Gold Run mine from dumping coarse debris into the North Fork of the American River. Descriptions of damage in surveys by Corps engineer Mendell (in 1875) and state engineer Hall (in 1880) had been cited in the trial. Judge Temple ruled that miners had never acquired the right to use rivers as dumps. Hydraulic mining constituted a public nuisance, according to Temple, because it blocked river navigation and the free use of riparian lands. At the same time, Temple wrote that he did not intend to block all mining. If the company built debris dams and convinced the court that these would safely restrain the coarse debris, the court would lift the injunction.

Both sides came to consider this the most important debris case to date. The Sutter County Board of Supervisors took the opening created by Judge Temple's ruling and filed suits for injunctions against every hydraulic mine operation on the Bear River, as Keyes had in 1876, and against several large mines on the Yuba that had contributed to flooding in Yuba City and Marysville.[45]

Edward Woodruff, who owned property around Marysville, also filed suit in the U.S. Circuit Court in San Francisco asking for a perpetual injunction against all mines on the Yuba River, including the vast North Bloomfield mine. That dispute had been brewing for some time. In 1881, the sheriff had shut off water to hydraulic mines in the area, and the mines had in turn sued to restart the water.[46] In 1883, Judge Lorenzo Sawyer rejected the argument that miners had used to defeat suits since the Keyes case. Sawyer ruled that because all the defendants were pouring debris into the river, there was no misjoinder of defendants. "The final injury is a single one . . . and all defendants cooperate in fact in producing it."[47]

The importance of the case became clear when Judge Sawyer appointed special commissioners to hear the expert testimony of nearly two thousand witnesses. Sawyer, who had been a forty-niner, personally inspected damage from debris, debris dams built by the state of California, and debris dams built by mine companies. He found that many dams were overfilled or broken. On 7 January 1884, Judge Sawyer took over three hours to read the verdict detailing damage to valley interests.[48]

Sawyer announced a permanent federal injunction against hydraulic mining by any company that failed to build restraints to hold all debris.[49] *Woodruff* was likely the first major ruling of a U.S. court to protect the environment. The ruling effectively ended hydraulic mining within the decade, although this was not foreseen at the time of the ruling.[50]

Debris from previous years of hydraulic mining, however, continued to flow into valley rivers for years, depositing sand as far as the San Francisco Bay. An estimated 1.3 billion cubic yards of mining debris entered tributaries of the Sacramento River.[51] Steamboat Slough had been the most important channel for steamboats on the Sacramento River. By 1880 it could not be used. In 1896, a Corps commission concluded it was "not worthy of improvement."[52] The California Commissioner of Fisheries reported in 1871 that debris had ruined salmon spawning grounds in the Feather, Yuba, and American rivers. Hydraulic mining had also permanently changed mountain excavation sites. In some locations, mining had cut so deeply into hillsides that it exposed bedrock in the form of steep cliffs. The North Bloomfield mine resembled a "barren amphitheater," according to one observer.[53] Debris persisted in the river channels well into the twentieth century.

Miners Organize for Aid

Politically active miners and legislators from mining districts began to work for aid to the mines. In 1888, they backed a bill by California representative Marion Biggs creating a Corps commission to study ways of reviving hydraulic mining without adding to farmland damage.[54] The Biggs Commission reported in 1891 that debris had caused forty-three thousand acres of farmland to be lost by 1880, causing a $2 million depreciation of land values. It concluded that building a system of debris dams would likely allow mining to restart under the conditions of the Woodruff ruling.[55]

In November 1891, miners from county mine associations met at Auburn, adopted resolutions expressing their grievances against the recent "history of oppression to the mining industry," and planned a state convention. Miners had been "long-suffering and slow to anger." It was time for them to "rise and assert their rights and organize in defense of their interests and their homes."[56] The *Mining and Scientific Press* argued that farmers had long complained that debris damage violated farmers' property rights.[57] Now, miners complained that the Woodruff ruling violated miners' property rights.[58]

The 1892 miners' convention in San Francisco attracted representatives from thirty-five mountain and valley counties. They formed the California Miners' Association as a permanent organization to revive hydraulic mining and sent a memorial to Congress backing the Biggs Commission's recommendations.[59] The Mining Congress, a national organization, had approved a similar proposal in December 1891, urging federal government to fund the debris dams as encouragement to an industry constrained by judicial rulings.[60] The San Francisco Board of Trade made donations to the California Miners' Association that were three times the amount donated by the largest hydraulic mining company. The California Miners' Association solicited resolutions supporting their memorial from the Alameda and San Francisco chambers of commerce and boards of trade, city and county supervisors of San Francisco and of mining counties, the Mining Congress, and the Trans-Mississippi Congress.[61]

Congress Creates the California Debris Commission

In 1892, Rep. Anthony Caminetti from the mining area of Amador County proposed forming the California Debris Commission as an agency in the Corps of Engineers. The commission would license mines that built debris dams meeting the requirements of the 1884 Sawyer ruling, and it would create flood and debris control plans. The California legislature and mining advocates sent petitions urging members of Congress to create the commission.[62]

Small and large mining companies split over the details of the bill. The California Miners' Association endorsed the general terms of the California Debris Commission bill in 1893. But owners of the largest mines opposed having the new commission regulate debris dams that they had already built in response to the Woodruff decision.[63] Residents who had disbanded the Anti-Debris Association of Sacramento Valley after the favorable Sawyer ruling, revived the association to protest the Caminetti bill. Negotiating for the Anti-Debris Association, former Rep. C. P. Berry got senators to add a penalty to the bill for violations of the bill's terms. The California Miners' Association protested the penalty clause, but Caminetti argued that protests would endanger the bill's passage.[64]

The proposal became law in 1893, with the proviso that California state provide matching funds.[65] The secretary of war appointed to the California Debris Commission the army engineers George H. Mendell, William H. H. Benyaurd, and William H. Heuer. All three men had been members of the 1888 Biggs Commission. The California legislature created the companion offices of State Debris Commissioner and State Commissioner of Public Works to work with the new federal commission.[66]

For years the California Debris Commission focused on reviewing applications for mining licenses. After monitoring the first meeting of the commission in 1894 and finding it trustworthy, the president of the Anti-Debris Association suggested that his association allow the commission to work without interference. Despite this, the *Mining and Scientific Press* complained that the Anti-Debris Association continued to file protests against licenses granted by the commission.[67]

Reclamation Work at the Turn of the Century

From this point on, most flood control activism in California centered on getting appropriations to fund the California Debris Commission's mission. Democrats were in control of the U.S. government, and the country suffering from the 1893 depression. Years passed before Congress appropriated funds for the commission that covered anything more than survey work.[68]

The drop in wheat prices that began in the 1880s also challenged farmers in the Central Valley. Lands changed hands, and wheat farming declined. In areas that could sustain intensive orchard or vineyard crops, and that had access to cheap labor, some owners divided and sold their large plots. A few farmers in the Sacramento-San Joaquin Delta established rice farms that were nearly as large as wheat farms had been in that area. Other delta owners were able to hold onto their large holdings by leasing their lands in small plots to Chinese, Japanese, Italian, and Portuguese tenants.[69]

Under the new State Commissioner of Public Works, A. H. Rose, California challenged Corps of Engineers orthodoxy. State engineers Marsden Manson and C. E. Grunsky designed a flood control system that rejected the single-channel policy still being promoted by the Corps of Engineers for the Sacramento and Mississippi rivers. Their system included debris dams in the mountains and levees in the valley, as did the Corps' plan. But they also proposed to enhance the natural bypass channels on the lower river, which drained water from the main channel when the river was unusually high. Valley resident and activist Will Green and State Engineer William H. Hall had presented early versions of this approach. A change in the governor's office, and the 1893 depression, ended their efforts.[70]

Over time, small farmers in the Anti-Debris Association became more sympathetic to the miners' call for aid. Navigation remained difficult above the city of Sacramento, where the Corps had not yet tried to make permanent improvements. River shipping declined through the 1890s, undercut by cheaper railroad rates. A flood in 1896 broke many debris dams and caused new flows into the valley. In response, the California Debris Commission proposed building more permanent debris dams.[71] The Anti-Debris Association sent delegates to Washington to plead for funding. A local congressmen secured $250,000 for a dam at Daguerre Point on the Yuba River, just upstream from Marysville, selected because that river held more debris than all the other tributaries combined. California matched this with funds it had previously approved.[72]

Most hydraulic mining had stopped after Judge Sawyer's ruling. Several of the largest mining operations had built debris dams and continued to work, in accord with the ruling. In some areas, smaller renegade operations continued to mine illegally for years.[73] A. J. Ralston, president of the Selby Smelting and Lead Company, was optimistic. In

1893, he commented that there was generally a better feeling about mining than at any time since the beginning of the "disastrous agitation" that led to Sawyer's 1884 ruling.[74]

Even so, the decline continued. Heavy snowfall in 1891 wrecked miles of water flumes in the mountains. Given the uncertainty whether they could meet Sawyer's requirements, plus high costs and declining productivity, many owners abandoned or sold their mines. Ditch companies had built six thousand miles of main ditches and over one thousand miles of subsidiary ditches by 1882. After the ruling, many abandoned or sold their water rights to the irrigation and power companies that now served the growing valley population. All of these moves signaled a shift in economic power toward agriculture and the cities.[75]

The California Debris Commission received few applications for mining licenses by the late 1890s, and it shifted its focus to river channel improvement. For members of the U.S. House Rivers and Harbors Committee, navigation remained the top concern. A delegation traveled to the region in 1901 to investigate whether the Daguerre Point dam was an expensive attempt to revive the mining industry that would fail to improve navigation. Pointedly, their boat stuck on a shoal. After visiting abandoned mining sites and meeting with valley residents, members decided that the dam was not primarily a miners' scheme. The 1902 Rivers and Harbors Act approved $150,000 for dams and channel clearing, later matched by California.[76] The Corps' navigation work continued in the 1910s with small appropriations. But it remained difficult to live with the river.

Conclusion

Conflicts in northern California among miners and farmers delayed the possibility of coordinated action by the California state government and the U.S. government. Congress, under the advice of California-based army engineers, did not attempt to intervene in flood control while local and state elites were still at odds about the federal government's role. Industrialization had set mine workers against mine owners and set small miners against large miners. Damage to farms caused by hydraulic mining had provoked splits between mining and farming capitalists and between mining and farming counties. Minor government interventions, prompted by the injured farmers, changed these conflicts only slightly over the years.

The 1884 Woodruff ruling in federal court essentially ended hydraulic mining. It soon changed the political field as well. Financiers, journalists, and commercial organizations in San Francisco depended on both mining and farming. They led the initiative to unify the major players in northern California. Once farmers had gained economic and legislative clout, a new political alliance became possible. Farmers who had earlier advocated an outright ban of mining became willing to join with other farmers and with miners to demand government flood control projects. Erstwhile opponents began work to socialize the costs of California mining to the entire country.

At the turn of the century, the lower Mississippi Valley was still dominated by labor-intensive agriculture, while California had built the groundwork for an industrial future. Federal aid was one element of the efforts by California elites to promote economic growth. It was a much more critical element in elites' plans for the lower Mississippi Valley. There, industrialization within the region was not sufficient to drive agricultural modernization or overall economic growth. Industrialization was still experienced as a pressure from outside that was making that region's products uncompetitive in the world market. For the Mississippi Valley, the cleavage between industrialists and landowners described by Stein Rokkan persisted.[77] In California, that cleavage was being overcome. Industrial processes were being integrated into California's mining, agricultural, and other sectors. The events outlined in this chapter and in chapter 6 overcame the political aspects of that cleavage.

State centralization was therefore experienced differently in the lower Mississippi Valley and in the Sacramento Valley. Even so, looking once again at the table at the beginning of part 2, we see that the Mississippi and Sacramento river areas produced markedly similar flood control campaigns. These two areas had the same property laws and the same federal system imposed upon them. Both also experienced the press of capitalism and industrialization. The point of the table is not to argue that specific structural conditions created regional flood control campaigns. Rather, it is to note that regional elites in two culturally and economically dissimilar regions responded politically in similar ways to local physical conditions and to the pressures of economic competition and political centralization. Faced with the emerging national market, local business leaders were motivated to develop land and rivers. They contended within their own regions to determine the objectives for resource development. A core of enthusiasts in these and other regions won passage of the Swamp Land Acts. That program created a group of politically savvy swampland owners eager to protect the most vulnerable low-lying lands. The existence of the Corps of Engineers' navigation program provided another opportunity. Activists demanded flood control surveys, plans, and, eventually, levee building on the Mississippi River and debris dams on tributaries of the Sacramento River. The next chapter discusses how the California activist movement came to resemble the movement in the Mississippi Valley. It also details how activists and legislators from both valleys joined their efforts and pressed for flood control bills.

part III

Redesigning Rivers in the National Interest

The expansion of central government activity during the New Deal marks the beginning of the fully bureaucratic state in the United States. The government's earlier activities regulating private access to resources and land, managing land-based resources, and building infrastructure projects set many of the preconditions that would allow the new bureaucracies to regulate the economy. In part III, chapters 8 and 9 discuss how activists and legislators got Congress to approve and then expand flood control projects for the Sacramento and Mississippi rivers before the New Deal. Chapter 10 outlines the rise of a national campaign that passed legislation during the New Deal, creating a nationwide flood control program.

The book concludes by discussing the organizational, legal, and political processes that led the government to intervene more deeply into land use. From the nineteenth century on, federal resource programs incorporated landowners and others with property-like rights into the processes of making and implementing policy. Land and resource programs also changed the relationship between state territory and private property and expanded the state's responsibility for intensifying land use as well as dealing with the ill effects of intensive use. These relationships and responsibilities continue to affect government relations with the citizenry, even though decisions about land have shifted from providing basic infrastructure and creating an agricultural base to building new infrastructure systems and facilitating a diversity of agricultural, industrial, and service uses.

chapter 8

Federal Aid for the Mississippi and Sacramento Rivers

> The river belongs to the Nation,
> The levee, they say, to the State;
> The Government runs navigation,
> The Commonwealth, though, pays the freight.
> Now, here is the problem that's heavy—
> Please, which is the right or the wrong—
> When the water runs over the levee,
> To whom does the river belong?
> —from Douglas Malloch,
> "Uncle Sam's River," in National Drainage
> Congress, *Official Proceedings*, 75.

River Activists Take Advantage of Government Expansion

By the turn of the century, activists seeking federal flood control aid for the Mississippi and Sacramento rivers were working alongside activists seeking federal flood control aid for all navigable rivers. Activists also adopted a new repertoire of political action by creating permanent membership organizations. These groups mobilized a coalition of Progressive Movement organizations and other groups, cultivated public sentiment, and lobbied government officials. Organizational work, personal politicking by activist leaders, and actions by sympathetic legislators won official flood control aid for the Mississippi and Sacramento rivers in 1917.

That activists had to work decades before they won large flood control projects points up politicians' great reluctance to create large federal programs. Debates about river policy between the 1870s and 1920s also evidence continuing, if muted, conflicts between the Northeast and the

South and West. Faced with these conditions, river activists in the Mississippi and Sacramento Valleys worked for incremental gains. They traded votes in Congress and broadened their public appeals during the Progressive period.

With the shift to permanent lobby organizations, navigation and flood control activists intensified their recruitment of supporters from around the country. Creating lobby organizations and calling for national river improvements—not just improvements for the Mississippi and Sacramento valleys—made it increasingly plausible for movement leaders to argue that specific river projects were in the national interest. Activists did focus on legislation for the Mississippi and Sacramento valleys in the 1910s, but they appear to have used calls for national flood control aid to build long-term support for flood control aid. The new lobby groups also used the rhetoric of efficiency and progress favored by the Progressive Movement. These changes in activist tactics appeared to distance the movement from its most obvious beneficiaries and suggested that a new flood control program would be run according to good government principles favored by Progressives. Provisions of the 1917 Flood Control Act requiring the Corps of Engineers to work closely with levee districts and with state government agencies did little, however, to change the localistic orientation of the Corps' earlier flood control and navigation work.

Permanent River Lobby Organizations Form —Mississippi Valley

By the 1880s, the federal government's reconstruction of the South had ended, rivers and harbors bills for navigation work had become a congressional mainstay, and army engineers had begun to build flood control levees on the lower Mississippi River. For decades following the disruptions of the Civil War, Republican politicians won national elections by reminding voters that Democrats had started the war. With their national dominance assured, the Republican Party could narrow its work to promoting northern business. Political entrepreneurs at work in the South slowly began to reshape Democratic Party objectives in statehouses and in Congress to emphasize government aid for southern development.[1]

From 1866 to 1882 Congress passed a rivers and harbors bill each year. The navigation bills had been favorites of legislators seeking to please local merchants, many of whom had turned to river shipping to avoid the railroad monopoly rates. During the years of the great industrial strikes and the rural populist activism in the 1880s and 1890s, this changed. Public critique of "the interests" that benefited from river aid and splits between flood control and navigation advocates in Congress resulted in rivers and harbors bills appropriating as little as $18 to $20 million every other year or so, then every third year.[2]

River activists responded to river aid cutbacks and to railroad competition by calling another round of conventions. They also formed permanent organizations to arrange conventions and undertake year-round mailings, organizing, and lobbying.[3] Flood control advocates had likewise tried for years to create a multistate organization of local levee districts on the lower Mississippi River. Planters and members of levee district boards from Louisiana and Mississippi finally met in Vicksburg in April 1890, while the river flooded, to form the Mississippi River Improvement Levee Association (MRILA).[4]

The MRILA president in the 1890s and 1900s was Charles L. Scott, perhaps the top planter in the Yazoo-Mississippi Delta. Scott was a lawyer who would later run for governor of Mississippi. His business associates included executives of the Illinois Central Railroad, which owned a rail line in the Delta, and other Delta planters. Planter John M. Parker, who later became governor of Louisiana, served as an officer of the MRILA.[5]

These planter-activists invited outside investors to the Delta and tied their own fortunes to national and international politics and capital markets.[6] Many of their fellow plantation owners were new to the business. In Louisiana's Sugar Bowl, it was estimated that half of the planters after 1870 were from the North or were supported by northern investments. Perhaps fewer than a third of the cotton plantations of the Mississippi Valley in the 1880s were owned by their wartime owners.[7] Scott, W. A. Percy, and Parker were transitional figures. Their slave-owning ancestors had risked fortunes to settle the swamps. Their descendants found it difficult to compete with northerners and Europeans who invested increasing amounts in bottomland real estate, agriculture, timber, and railroads in the South.

Planters, farmers, and engineers in the MRILA sought to boost southern economic development and to counter growing local criticism that spending on levees was wasted. They published detailed reports of successful local levee projects.[8] Like the Corps of Engineers and the Merchants' Exchange of St. Louis, the MRILA opposed outlets and promoted levee building as the sole viable solution to flooding.[9] An additional $10 million from the federal government would, according to the MRILA, " 'complete a system of levees that will prove entirely effective to restrain all future floods.' "[10]

At its founding convention in 1890, the MRILA reported that a flood on the lower Mississippi River in 1882 caused 284 crevasses in the levee system. Because the levees had been improved, however, a more intense flood in 1890 resulted in only twenty-three crevasses. A congressional representative from Tennessee attending the 1890 convention asserted that levee building would help the freedmen, who were destined to grow cotton on the lowlands "for all time to come."[11] The Merchants' Exchange of St. Louis, Southern Press Association, New Orleans Board of Trade, Commercial Club of Louisville, Mississippi and Ohio River Pilots' Society, and a committee of engineers from the army and from levee districts sent resolutions to the convention in favor of federal aid.[12] MRILA officers established a Bureau of Correspondence in Washington, D.C., to create propaganda and to lobby Congress for flood control funds, and they set an 1891 convention for Greenville, Mississippi. The organization also circulated thousands of documents around the country and sent speakers to other river organizations to testify to the value of levees.[13]

Congress approved $1 million in emergency aid to rebuild Mississippi River levees after the 1890 flood. The previous year, the break of a neglected dam at a private resort owned by Pittsburgh industrialists had killed over 2,200 in the working-class town of Johnston, Pennsylvania, making the public sympathetic to flood victims.[14] This sympathy was politically important. Many members of Congress still considered levees ineffective, and others opposed federal involvement on principle. Until then, federal navigation improvements bills had typically stated, "no portion of this appropriation shall be expended to repair or build levees for the purpose of reclaiming lands or preventing injury to lands or private property by overflow." But during hearings on the emergency bill, legislators discussed direct aid for flood control. The clause in the final bill was rewritten to read, "for the general improvement of the river, for the building of levees . . . in such a manner as in [the Mississippi River Commission engineers'] opinion shall best promote the interest of commerce and navigation."[15] The MRILA characterized this bill as "first distinct triumph of levee interests."[16]

The federal Mississippi River Commission reported that it had spent more on levees during the mid-1890s than local and state sources had spent. Commission reports also acknowledged that while the improved levee system appeared to have increased river volume and height, contrary to earlier expectations of many commission members, increased protection for riverfront lands justified the policy. Levee failures in an 1897 flood prompted a congressional investigation, but legislators were impressed by a pamphlet by Yazoo levee engineer William Starling—who was affiliated with the MRILA—affirming the growing inclination of the Mississippi River Commission to close outlets and build higher levees, and they recommended continuing to build levees.[17]

In a subcommittee hearing of the Senate Commerce Committee in 1897, senators requested that the MRILA provide the subcommittee with data on the alluvial areas and their economies. The MRILA also continued producing propaganda and making contacts with commercial organizations around the country. For example, the MRILA circulated a booklet of reprints from the *Manufacturer's Record* detailing investment opportunities in the South. The booklet promoted their version of the history and culture of the lower Mississippi, showing photos of white men on bear hunts and of an African American child eating watermelon. It also advocated development of the region's levees. In that booklet, Louisiana Supreme Court judge N. C. Blanchard remarked: "While somewhat of a strict constructionist of the Constitution, I have no trouble in deducing authority from that instrument justifying Congress in appropriating funds for the two-fold purpose of improving the navigation of the river and of restraining its flood waters."

"This has been disparaged by some as a lobby or lobbying plan, but what if it is?" wrote MRILA special representative Frank Tompkins. Such an organization is not a problem, according to Tompkins, if it is used properly.[18] Congress presented $5 million to the Mississippi River Commission for levee building in 1897. With the 1903 meeting in New

Orleans attracting over 1,000 delegates from 27 states and 166 cities, the MRILA became the first strong association of levee districts to lobby for federal aid.[19]

Permanent River Lobby Organizations Form —Sacramento Valley

During this same period, activists in California also organized permanent groups to pursue similar ends for the Sacramento Valley. The formation of permanent lobbies in both valleys facilitated activists' coordination on the national level. It also facilitated cooperation among southern and western members of Congress who traded votes on water resource issues.

The economic potential of California agriculture was evident by the turn of the century, and control over water seemed the key.[20] In May 1902, people interested in California's one million acres of overflowed, little-used swamplands called a mass meeting in Sacramento. They voted to form the Reclamation and Drainage Association, renamed the River Improvement and Drainage Association of California. The bankers, landowners, and industrialists attending from the Sacramento and San Francisco areas knew that lawmakers were loathe to fund river projects perceived to benefit landowners. They hoped to make clear the limits of private actions to reclaim land from swamps: "You can not improve the rivers without helping reclamation; on the other hand, a world of money can be spent in reclamation without improving the rivers."[21] The developed areas of northern California accounted for three-fourths of the state's assessed valuation, they noted. The public interest would therefore be served if rivers in the Sacramento and San Joaquin valleys and in the San Francisco Bay counties were straightened and deepened. Levees would also have to be increased, they argued, and weirs would be added to control dangerous surplus waters.

Development of the Sacramento River to date had been piecemeal. Small reclamation districts had built levees on the west bank of the Sacramento River to block natural outlets and protect local lands. These levees had also raised the river's flood plane, in effect gaining protection for upstream areas by drowning their downstream neighbors.[22] So, the association urged that studies by the California Commissioner of

Public Works begun in 1897 be expanded and that a comprehensive state government plan for reclamation and navigation be developed.[23] Getting federal funds approved would inevitably involve delays, and so the executive committee recommended asking the state to provide most of the funding for new works.[24] In consultation with Corps officers and state government engineers, an executive committee for the convention drew up a levees-only flood control plan. This $750,000 proposal failed in the California legislature.[25]

Hoping to overcome political and economic rivalries between northern and southern California, civic elites in San Francisco formed the Commonwealth Club in 1903. The club would investigate and promote Progressive planning efforts on a wide array of social needs, including river development. The club heard from state and federal government engineers who had worked on the Mississippi and the Sacramento rivers.[26]

A lead organizer of the club and editor of the *San Francisco Chronicle*, E. D. Adams, did detailed research on the history of Sacramento Valley flood control efforts and recommended that the state of California undertake a centralized program of reclamation. Adams cited a court ruling on a dispute over a Swampland Grant claim. The judge declared that the claimant had bought the land from the state government with the understanding that the state government would then reclaim the land. Adams argued further that the region also deserved federal aid because California gold mines had helped the United States resume specie payment at the end of the Greenback period and had helped finance the Civil War. Adams and other club members cited the levee work that Congress had approved for the Mississippi River, which had clearly benefited private landowners. Adams noted that many in California saw this as a possible precedent for similar aid to their own Sacramento River. Commonwealth Club members studied the options of taxation versus bond issues. They debated whether landowners or a broader range of beneficiaries should pay for the improvements. Then they adopted resolutions calling for the state government to take the lead in organizing reclamation, as the federal government seemed unlikely to do so in the near future.[27]

Floods in 1903 and 1904 inundated several of the most securely leveed reclamation districts on the Sacramento and San Joaquin rivers. The damage prompted calls for former mining and farming adversaries to unite. The California State Board of Trade, the Commonwealth Club, and thirteen merchants' exchanges and chambers of commerce called for a convention to consider navigation and overflow problems in the Sacramento and Stockton areas. Editor E. D. Adams promoted the idea in the *San Francisco Chronicle*.[28]

In May 1904, over four hundred rural landowners and members of the urban elite met in San Francisco. They established a statewide lobby under the same name adopted by the 1902 Sacramento convention, the River Improvement and Drainage Association of California. Delegates included Gov. George Pardee, Sen. George C. Perkins, and representatives from boards of supervisors, local trade associations, counties, and reclamation districts. Delegates noted that the Corps of Engineers had already expanded its work to include more than navigation improvement. They resolved to press the state government to increase its work on rivers and to work with property owners and agencies at all levels of government.[29]

The Director of the California Board of Trade, William Harrison Mills, asserted that without government improvement of river control works, the river channels would become filled with mining debris once again.[30] Convention delegates proposed that state, federal, and civil engineers plan a flood control project funded by landowners, California, and the federal government. Convention delegates created a permanent committee to continue the convention's work, which merged with a committee of the Sacramento Valley Development Association. The *San Francisco Chronicle* reported that "discussions resolved themselves into a big love feast for the good of the great interior valleys."[31]

California's governor followed the recommendations of the River Improvement and Drainage Association of California by creating a Board of River Engineers in 1904. He approved the three members chosen by the association, engineers who were well versed in methods used on the lower Mississippi River and experienced in dealing with the demands of landowners. The commission included T. G. Dabney, a Corps officer and the manager of the Yazoo-Mississippi Delta Levee District; Henry B. Richardson, the Chief State Engineer of Louisiana and a twenty-four-year member of the Mississippi River Commission; and H. M. Chittenden, a Corps officer who had worked on the Missouri, Ohio, and upper Mississippi rivers.[32]

The engineers recommended that until the full channel depth was achieved, the state government should relieve pressure on levees by building escapement weirs to allow excess water to temporarily bypass the channel. The board rejected the proposals by California engineers Marsden Manson and C. E. Grunsky to build permanent bypasses and recommended a system dominated by levees instead. Based on the approval of the California Commissioner of Public Works and the River Improvement and Drainage Association, the California legislature adopted this plan.[33] Negotiations over the plan's provisions would continue for several years in California and in Congress.

National Politics and Land at Century's End

Northeastern Republicans who dominated the presidency and Congress had been stingy in meeting demands from the South and West. In response, river development activists began mimicking the rhetoric of business-minded and Progressive Republicans, including arguments against agrarian populism.[34] Partisans of the Corps of Engineers talked of the efficient use of resources. This attracted President Theodore Roosevelt to their cause, even though Roosevelt's closest Progressive ally in conservation, Gifford Pinchot, waged a long battle against the Corps precisely because he considered it inefficient.

Agrarian populist protests against capitalist institutions in the West and South had flourished in the 1890s as the country entered its worst depression to date but did not block the progress of river developers. Smallholders near and along the Mississippi and Sacramento rivers formed populist alliances, and a populist movement in California helped to force the state constitutional crisis (discussed in chapter 7). Populists failed to achieve their key goals, however. Federal policies continued to limit credit to small farmers. Largeholders continued to consolidate failed farms in most regions. And populists were unable to establish permanent alternative credit and marketing institutions.[35] Largeholders viewed levee building as one means of maintaining their political and economic ascendancy over smallholders and laborers. By buying most or all of their holdings within a given district, providing steady trade to local contractors, participating in local politics, and framing local growth issues to their own advantage, owners of large holdings in the Mississippi and Sacramento river valleys gained local power and held off populism locally.

The cause of Mississippi River development was furthered by the institutional power that white southern Democrats built in Congress, even under Republican leadership. By blocking most freedmen from voting, and by cowing many poor white voters, southern Democrats established many safe seats in Congress. Under the seniority system, those in safe seats gained a disproportionate share of committee chairs. Their longevity also made it easier to vote in blocs.[36] During congressional hearings in 1922, it was acknowledged that members of Congress from the South had often traded votes with other members—even at the sacrifice of other southern interests—to secure votes to pass flood control bills.[37] Frequently this involved trading votes with members from the Far West on irrigation and other land development bills.[38]

Into this political mix, Theodore Roosevelt (Republican, 1901–9) was elected president. Roosevelt favored developing the trans-Appalachian West in cooperation with large-scale landowners and extraction companies, as long as they used resources rationally. Roosevelt expanded national monuments and parks, withdrew mineral lands from immediate sale, and increased conservation enforcement. He also urged the army to foment an "uprising" to make it possible to build the Panama Canal and set out to rationalize domestic river management.[39] As the Democratic Party regrouped, and as party discipline declined, Roosevelt needed the help of interest groups to pass his land policies.[40] River activists attracted support from Roosevelt and other Progressives by insisting that they intended to replace pork barrel river politics with rational decision-making processes.

Beginnings of the "Lobby That Can't Be Licked"[41]

Some activists and engineers continued to disseminate information about flooding and river improvements to social and civic organizations.[42] Others began to build a member-based lobbying organization of the form that would characterize twentieth century pressure-politics. In protest of the 1901 filibuster of a rivers and harbors bill, leading New Orleans shippers and bankers brought together several local river organizations from around the country to form the National Rivers and Harbors Congress (NRHC).[43] Adopting the slogan, "a policy, not a project," the NRHC sought to dispel the image of the river program as a pork barrel venture bloated by congressional logrolling. They would instead support river projects approved by the Corps in consultation with local leaders.[44] The chair of the House Rivers and Harbors Committee, Theodore E. Burton, passed a bill in 1902 favoring this NRHC goal. Burton's bill created the Army Board of Engineers for Rivers and Harbors, which would assess proposed river projects and thus discourage members of Congress from adding pet projects.[45] The NRHC was the first organization to call for a nationwide program of federal river development. It was also the first to consult directly with congressional committees and with the Corps of Engineers.[46]

The NRHC held each annual meeting in Washington, D.C., and moved its headquarters there in 1911, making clear their interest in lobbying. In an effort to make the NRHC a truly national organization, Sen. Joseph E. Ransdell (Louisiana), chair of the NRHC Executive Committee, traveled with other legislators in 1906 to meetings of commercial organizations and waterway associations around the country. The NRHC's Publicity Bureau sent letters to newspapers in every state announcing the organization's expansion efforts. The NRHC meeting that year opened with a speech by the powerful Speaker of the House, Joe Cannon, in support of bills passed by the Rivers and Harbors Committee.[47]

Representatives from shipping companies, regional trade and river development organizations, chambers of commerce, farming organizations, and levee districts from many river valleys joined the NRHC. The 1906 convention drew representatives from 189 commercial associations. In 1909, the men's organization had 6,300 members and the Women's National Rivers and Harbors Congress had 2,500 members. Their conventions consistently attracted many congressional representatives, senators, Corps officials, governors, and cabinet members as delegates or speakers. There were fourteen governors at the 1906 convention and 287 mayors at the 1908 convention.[48]

Members organized speakers bureaus and sent representatives to many cities to speak about river development.[49] At a meeting of the California Counties Committee in 1909, the special director of the NRHC, John A. Fox, recommended lobbying Congress for $50 million a year for rivers.[50] The women's NRHC organized mass mailings of circulars and other material; gave lectures to women's and men's clubs, teachers' conventions, and civic organizations; introduced waterways into school curricula in many cities; and prompted the General Federation of Women's Clubs to create a waterway committee. By the 1910s, Rep. Stephen M. Sparkman, later chair of the House Rivers and Harbors Committee, declared that the rivers and harbors bill depended much on the NRHC's activities.[51]

California Activists Make National Links

Northern California river activists began to work with the NRHC when it reinvigorated its national outreach in 1906,[52] and they continued to organize within California and to lobby Congress for Sacramento River projects. River Improvement and Drainage Association representatives sent a resolution to Congress in 1905 calling for a coordinated plan for developing the Sacramento River system. The plan would include a valley-wide drainage district with a board elected by landowners, professional land assessment for rate charges, a three-way split of costs between owners and the state and federal governments, and a board of river control to oversee river projects. The state government and the Corps began this phase of work by installing a complete system of water gauges on the Sacramento and San Joaquin rivers.[53]

The former California State Engineer, William Hamilton Hall, presented a report to the Corps of Engineers asserting that the general government had peculiar responsibility for repairing damages from mining. "A great National-encouraged industry, whereby the Nation was enriched, in its prosecution debauched the main waterways of this commonwealth." The government had utterly lacked an adequate system to regulate and protect the rivers, he argued. The Corps of Engineers had failed to forecast the effects of mining. Furthermore, structures that the Corps had since built to improve low-water navigation impeded the ability of the river to carry flood-stage volumes. Even if the state or individuals had the means to improve the river's capacity for floods, they could not remove these national government works. "This the government should have foreseen and prevented, or have assumed responsibility for the consequences."[54] Hall urged federal action.

Most members of the federal government's California Debris Commission had been doing flood control studies on the Sacramento Valley since the 1890s and were sympathetic to local interests.[55] In 1902, the commission recommended that structures be built on the Yuba at an estimated cost of $800,000 to settle out mining debris to prevent it from clogging the Sacramento. Seeing insufficient agreement among California residents and investors, however, the commission recommended in 1905 that the federal government merely dredge and levee the lower Sacramento River until landowners and the state government agreed to a common plan.[56] The U.S. House Rivers and Harbors Committee, chaired by Representative Burton, likewise refused to extend the Corps' duties on the Sacramento beyond navigation work.[57]

Once again a mammoth flood spurred action. Floods in 1907 on the Sacramento and its tributaries destroyed the mountain dams that the California Debris Commission had built in the 1890s and destroyed many levees. Engineers had estimated normal flow on the Sacramento River to be 6,000 cubic feet per second and a maximum to be 300,000 cubic feet per second. State government gauges registered a flow of 600,000 cubic feet per second during the flood.[58]

The California Debris Commission used increased public interest to press Congress in 1907 for more funding.[59] Congress appropriated $400,000 to build two dredges, and California matched this amount. The commission also began to design a new flood control plan for the Sacramento Valley.[60] The River Improvement and Drainage Association convinced the Army Board for Rivers and Harbors, the new oversight board, to tour the Sacramento River and to attend open hearings in the valley. The Army Board of Rivers and Harbors and Acting War Secretary Robert S. Oliver recommended in 1908 that the first comprehensive survey be undertaken.[61] The River Improvement and Drainage Association and regional newspapers encouraged Californians to write in support of the bill.[62] Once members of the House Rivers and Harbors Committee understood that landowners, not navigation interests, were behind the California plan, they rejected the plan.[63]

Rationalized Rivers

Theodore Roosevelt supported and promoted the objectives of navigation and flood control activists, who exerted political pressure behind the scenes and as members of lobby groups. Roosevelt was a friend of John M. Parker, an officer of the MRILA who became governor of Louisiana. Parker introduced Roosevelt to other Yazoo-Mississippi Delta planters, including LeRoy Percy, who became the leading flood control activist among planters.[64] Serious flooding on the lower Mississippi River in 1897 and in 1903 had inspired these and other planters to undertake more active campaigning for flood control projects, through the MRILA and through their personal contacts.[65] Parker and Percy began to pass their political requests to the president via House Speaker "Uncle" Joe Cannon.[66]

During Roosevelt's terms as vice president and then president, he supported river activism and helped to organize the NRHC.[67] Roosevelt attended a 1907 river convention in Memphis with LeRoy Percy, who had become Roosevelt's friend. There, Roosevelt gave a speech in favor of multiple-use river development, including flood control.[68] Over the years, Roosevelt praised the NRHC in speeches as the only organization favoring a general program for river development instead of promoting specific projects to benefit its members. Convention delegates from thirty-two states visited President Roosevelt en masse in 1906, and in 1907 delegates visited him and presented resolutions. But even with Roosevelt's pressure, Congress refused to fund river bills with any consistency.[69]

Roosevelt and Gifford Pinchot wanted the Corps of Engineers and river activists to back a broader program of conservation by supporting multiple-purpose water and hydroelectric development.[70] In speeches before the NRHC, Pinchot and irrigation enthusiasts, such as William McGee and Sen. Francis Newlands of Nevada, promoted multipurpose development by the Corps and other agencies.[71]

Pinchot, McGee, and irrigation activist Frederick Newell meanwhile convinced Senator Newlands and President Roosevelt to support a new Inland Waterways Commission. This commission of administration officials would design integrated water resource plans outside the control of the House Rivers and Harbors Committee and its favorite agency, the Corps of Engineers.[72] The new commission recommended creating a strong, permanent central conservation agency, which Roosevelt favored. Chief Army Engineer Alexander Mackenzie was the sole dissenting member. Corps officers and river activists lobbied against making the Inland Waterways Commission a permanent body outside the control of the Corps, and private power companies opposed it for its potential to regulate hydropower generation. Members of Congress, who themselves were being asked to relinquish control to an independent agency, rejected several bills over the years.[73]

Even so, other form of multipurpose development did begin to erode the Corps's monopoly on river policy. Roosevelt expanded the national forest system in part to protect watershed forests as natural reservoirs. He believed headwater forests could store enough water to reduce river flooding. Roosevelt considered this policy to be constitutional because it was similar to building flood control levees on the Mississippi River, even though Congress had not yet officially recognized levee work for flood control as constitutional.[74] In the 1909 Rivers and Harbors Act, Congress ordered the Corps itself to study the potential for hydropower development at its project sites to help subsidize the Corps' other work.[75] Managers of the Reclamation Service embraced the multipurpose philosophy enthusiastically, using hydropower as a means of subsidizing the Reclamation Service's irrigation work and of enabling that agency to challenge the Corps' authority on certain rivers.

Bolstered by the support of merchants in many localities, Corps officials resisted congressional proposals to expand beyond its navigation improvement duties to build hydroelectric facilities, and they opposed Reclamation Service proposals to build upstream reservoirs, arguing that such projects could harm navigation. Publicly, some officers even continued to argue against undertaking flood control as an official duty, even though Corps units were already doing such work on the Mississippi.[76] Not until 1912 did a Mississippi River Commission report officially admit that its primary objective in building levees was to protect land from overflow.[77] At the 1911 NRHC convention, the Corps' Chief Engineer W. H. Bixby stated that Corps engineers recognized the need for multipurpose developments, including power, irrigation, drainage, bank and levee protection (revetments), and levee building.[78]

California Starts the Sacramento Flood Control Plan

In 1909, California boosters formed the San Joaquin and Sacramento River Improvement Association to overcome the conflicting local interests that were delaying action. The River Improvement and Drainage Association, Sacramento Valley Development Association, Sacramento Drainage Commission, California Miners' Association, and irrigation interests encouraged property owners to join their efforts in the new association. The Sacramento River Improvement Association formalized interconnections among leaders of activist organizations.[79] Once the interested parties could agree on objectives, it would "simply [be] a matter of sufficient money to do the necessary work," according to one mountain paper's editorial.[80] This new association proposed to work with army engineers and other government authorities to secure rights-of-way for private and public flood relief and land reclamation projects. They backed the California Debris Commission's plan to straighten and widen the Sacramento River below the Sacramento-San Joaquin Delta. They also proposed to raise $235,000 to meet the cost-share that local interests were required to provide before $800,000 in state and federal funds could be spent on the river.[81]

River development became complicated as competition for water increased in the valley, but opposition from miners had evaporated. Drift mining sources had became exhausted, and hydraulic mining proved uneconomical under the California Debris Commission's requirements. Total gold production was below $1 million a year by 1908.[82] Even newspapers and chambers of commerce in former mining towns argued for improved river control and reclamation.[83] A report of the Commonwealth Club in 1909 remarked upon the great increase of investment in the Sacramento Valley.[84] About 300,000 acres in the entire Central Valley were reclaimed by 1910.[85]

California Debris Commission member Thomas H. Jackson began to design a new plan rejecting the levees-only approach. His plan included new mountain dams, levees, and, most importantly, bypasses alongside most of the leveed portion of the river. The bypasses would replicate the historical flow of occasional excess floodwaters away from the river's main channel. The federal Board of Engineers for Rivers and Harbors decided that the plan would be financially acceptable if California shared costs but that it was not essential to navigation needs. Congress therefore authorized work only to increase the capacity of the river downstream from Cache Slough.[86]

Progressive Gov. Hiram Johnson called a special session of the California Legislature in 1911 to approve the state's participation in Jackson's Sacramento Flood Control Plan. He also proposed a state Reclamation Board to coordinate reclamation, flood control, and navigation projects with the federal government.[87] Local levee districts had fostered "a struggle or war, in which the biggest and strongest levee would certainly be the winner," according to the state engineer. The new plan therefore allowed the Reclamation Board to prosecute owners of reclamation works that did not conform to state flood control plans.[88] In 1913, the legislature also created the new Sacramento and San Joaquin Drainage District to coordinate reclamation activities of swamplands in the Central Valley. The district covered about 1.5 million acres administered by 120 government agencies and thirty-seven private reclamation companies.[89] Some landowners acquiesced to the state Reclamation Board only after having their day in court. Land reclamation in the Sacramento-San Joaquin Delta soon, however, produced an agricultural boom in asparagus, beans, and celery, farmed mostly by tenants.[90] Congress did not fund federal participation in the California Debris Commission's plan.

A River Policy Issue Network Forms

In determining direct improvements for inland navigation, the Corps of Engineers, the NRHC, the House Rivers and Harbors Committee and, to a lesser extent, the Senate Commerce Committee by this time had constituted themselves as an issue network, working in close association to shape river policy and facing effective resistance only on occasion. President William Howard Taft (Republican, 1909–13), from outside this dedicated group, paid some attention to the issue, speaking at several NRHC conventions. At a convention of the Lakes-to-the-Gulf Deep Waterway Association, he called the NRHC the leading waterway association framing broad policies on rivers.[91] Taft created the National Waterways Commission in 1912 to guide multipurpose resource development with Sen. Theodore E. Burton as its chair. Burton was the former head of the Inland Waterways Commission and former chair of the House Rivers and Harbors Committee, and he preferred maintaining the Corps' dominance over river work, rather than constituting an expert commission to impose comprehensive planning from outside the agency.[92] Burton also defended the rivers and harbors bills against critics, remarking: "There may be a little bit of log-rolling, but I challenge anyone to cite an instance where that bill has ever been made up to gratify certain localities or to advance the interests of some member of Congress."[93]

Burton and other officials who determined the scope, funding, and implementation of federal river improvements dutifully appeared before the NRHC. Federal officials attending included members of the Mississippi River Commission, officers of the Corps of Engineers, secretaries of war, and chief army engineers.[94] Reforms led by Progressive Sen. George Norris of Nebraska had reduced the power of the Speaker of the House in 1910, and as discretion over most bills became vested in the committees, the river issue network had begun to form.[95] Confirming the vital role of the NRHC were appearances at its conventions by several Speakers of the House and by members and chairpersons of the Senate Commerce Committee and of the House Rivers and Harbors Committee.[96]

Before the NRHC was founded in 1901, Congress passed rivers and harbors bills every three years with only $18 to $20 million in appropriations for navigation. With satisfaction, the NRHC's president, Sen. Joseph E. Ransdell (1907–19), observed in 1911 that Congress was now passing bills with $30 million. At the 1911 convention, the NRHC set a new goal for $50 million in each year's bill.[97]

Activist Priorities Shift

Simultaneous moderate floods on the Missouri, Ohio, and upper Mississippi rivers combined into a major flood on the lower Mississippi River in 1912 and a second flood in 1913, which received heavy play in newspapers around the country.[98] The 1912 flood killed over 2,000 in Ohio. Even though the water stopped just below the emergency sandbags on top of levees in the upper Yazoo-Mississippi Delta, these floods ended confidence that work on Delta levees had been finished. The

Mississippi River Commission's postflood recommendation for larger levees motivated local levee districts to agitate harder for federal aid to meet the expected costs.[99]

The NRHC had focused on lobbying for river navigation improvements. After the 1912 floods, its delegates began to urge that the organization devote more attention to promoting flood control. John H. Small, a member of Congress from North Carolina, and later NRHC president, acknowledged that Congress gave the flood victims relief aid but that they needed continued "immunity." Small and subsequent speakers argued that Congress should assert plainly that flood control is justified because of the valley's great resources.[100] J. Hampton Moore, a member of Congress from Pennsylvania and president of the Atlantic Deeper Waterways Association responded hotly. Moore warned that the NRHC should not "lose hold on the main chance and tie up our great fight to every bob that comes along."[101] His was a minority view in an organization with many members from the lower Mississippi Valley.

On the Mississippi and Sacramento rivers, the Corps' work directly involved levee district officials and farmers. Since the 1880s, it had been difficult to distinguish flood control from navigation work on these rivers. For instance, at one point Congress cut funding for mats to protect levees from wave erosion, devices that had been approved for federal funding because they benefited navigation as well as flood control. In response, the upper Yazoo-Mississippi Delta levee district funded Corps crews to continue adding these revetments to preserve levees protecting district lands.[102] Such connections between navigation and flood control work may have made it difficult for the NRHC to avoid supporting flood control once a group of NRHC members demanded it.

Other organizations also took up the call in response to the floods. The Louisiana Engineering Society, Progressive Union, Contractors' and Builders' Exchange, and Louisiana State Medical Society all called for federal aid.[103] Speaking before the Lakes-to-the-Gulf Deep Waterway Association, the general manager of the Illinois Central Railroad stated that the government must ask if it is responsible for the flooding, and it must answer "yes."[104]

With the flood year of 1912 being a presidential election year as well, the political opportunities were evident. In the Rivers and Harbors Act of 1912, legislators directed the Mississippi River Commission to buy private lands that were too expensive to protect with levees but were regularly flooded due to the placement of other levees.[105] President Taft declared that the devastating flood made it obvious that reclaiming low-lying lands on the lower Mississippi River is a national problem.[106] His party's 1912 platform stated that the federal government should assist states to build levees and should assume a "fair portion" of the burden. Theodore Roosevelt's renegade Bull Moose party supported forest development, reservoirs, and levees. Woodrow Wilson's Democratic Party platform included a plank written by Senator Ransdell supporting a full federal levee program on the lower Mississippi River.[107]

Even so, the new chair of the Rivers and Harbors Committee, Stephen Sparkman of Florida, rejected the idea of river work for purposes other than navigation. He sent the California Debris Commission's river plan back to the chief of engineers with instructions to strip out projects that provided no benefit to navigation, namely the flood bypasses.[108]

Growing Grassroots

Flood control activist organizations created a network of supportive civic groups and commercial organizations to project the appearance of—and to some extent to create—broad public support. Shippers, local officials, and members of levee district boards remained politically active but became less visible and focused on working directly with administration officials and members of congressional subcommittees.

A convention of planters and levee district officials in Memphis in 1912 created the Tri-state Levee Association, later enlarged and renamed as the Mississippi River Levee Association (MRLA). This soon became the most powerful flood control lobby.[109] Even before the 1912 flood, planters in the upper and lower Yazoo-Mississippi Delta levee districts had started raising money for a publicity campaign and arranged for the new association's organizing meeting.[110] Planter Charles Scott—the president of the first association of levee districts, the MRILA—and planter LeRoy Percy became board members of the MRLA. They continued to lobby their Washington, D.C., acquaintances.[111] Former president Teddy Roosevelt gave an enthusiastic speech at the 1912 MRLA convention in favor of aid for flood control.[112]

The board of the MRLA (see appendix 2) included well-connected bankers, wholesalers, railway officials, editors, lawyers, manufacturers, merchants, planters, and levee district officials, mostly from the lower Mississippi.[113] Thousands of business people in the valley contributed to the MRLA campaign. Eight of the leading regional railways and other leading regional businesses gave the association $1,000 per year for several years. Members gathered data about river flooding. Using these data, MRLA secretary-manager John A. Fox wrote thousands of letters, dozens of pamphlets, and several books. Fox distributed pictures, films, maps, and charts and organized a team to make thousands of lectures. The MRLA appealed to northern whites' sympathy for African Americans in the South by featuring in its publications photos of black farm laborers and tenants stranded in the 1912 flood.[114]

Fox sent special trains full of southern businessmen to Washington, D.C., for long visits to lobby Congress. He organized letter-writing campaigns by state and national river improvement conventions, by bankers, by mayors of large and small cities, and by associations of commercial and civic organizations. Speaking before a NRHC convention, A. S. Caldwell, president of the MRLA, said that spending $60 million for reclamation on the lower Mississippi River would gain sixteen million acres "without firing a shot" and would perpetuate the country's cotton supremacy.[115] The MRLA's glossy publications reprinted letters of endorsement from supporters and included dramatic photos of emergency work crews and of flood damage.[116]

"We ask for protection, not reclamation," Fox asserted. Federal flood control aid would enable private landowners to devote their own resources to reclaim lands near the river and increase food production. "Reclamation in the delta will follow protection just as civilization followed military occupation in developing the virgin West."[117] The MRLA nonetheless drew criticism that it was a lobby organization seeking taxpayers' money to support special interests.[118]

The National Drainage Congress (NDC) rejected the cautious approach of the MRLA and defied those who criticized subsidies for private land development. The NDC was organized in 1911 to demand not only flood control aid but also federal aid for draining "overflowed" private lands. Some NDC convention speakers argued that the general welfare clause of the Constitution justified drainage, in part because malaria still occurred in swamp areas. They also made the dubious suggestion that canals draining off groundwater and rainwater could be used for navigation.[119]

The NDC aimed to organize a large membership including elected officials, delegates appointed by governors and mayors, and representatives from resource commissions and state engineering agencies. Most NDC conventions met in the Ohio and Mississippi river valleys, although conventions also met on the East Coast and in California. Two flood control engineers prominent in California water politics, V. S. McClatchy and C. E. Grunsky, attended the 1916 meeting in Illinois. Although speakers to the NDC discussed drainage issues in many parts of the country, they emphasized the potential value of drained swamplands on the Mississippi River and in California. The NDC sent delegates to lobby Congress and kept a representative in Washington, D.C. A committee from the NDC went to each party's presidential nominating convention in 1916 to influence party platforms. Members consulted with officials and staff members of national resource and conservation agencies who were writing drainage bills. The second NDC convention, which took place during the 1916 flood of the Mississippi River, attracted one thousand delegates from thirty states.[120]

In weighing their chances for support from federal agencies, NDC officials chose to advocate multipurpose river planning and development. They cultivated connections with the secretary of the interior rather than the chief army engineer.[121] In later legislative battles, NDC officials would work against the aims of those seeking to build the Corps' flood control program.

Conflicts over Resource Policies

During Democrat Woodrow Wilson's terms as president (1913–21), land agencies began to compete vigorously for funds and for authority over natural resources. Most local activists who sought river development aid chose to ally with the NRHC and the Corps of Engineers. Some, especially from the arid and mountainous areas of the Far West, saw advantages in advocating Pinchot's philosophy of multipurpose development and in linking their fortunes to the federal government's Forest Service or Reclamation Service.[122]

In the Sacramento Valley, the Reclamation Service began to attract local support for its plans to build upstream reservoirs. These plans were especially attractive in the 1920s, when droughts and water diversions by private irrigation companies reduced water supplies for others. Corps officers eventually decided as a general policy that such reservoirs might alleviate some of the competition over water.[123] Flood control activism by advocates of the Corps of Engineers occurred against this background of ideological and bureaucratic conflicts.

As the river lobby became more prominent, criticism of the Corps increased.[124] Most of it was the self-serving criticism of those who favored other resource agencies. Some critics were genuine advocates of clean government. Sen. Francis Newlands, author of the 1902 irrigation act, consistently criticized Corps management. From 1909 to 1916, Newlands and Sen. Edwin Broussard of Louisiana introduced bills for comprehensive river development, including bills to make Theodore Roosevelt's Inland Waterways Commission a permanent agency.[125] Andrew Carnegie had attended two earlier NRHC conventions, but he and other prominent figures now criticized the NRHC's influence. Sen. Theodore E. Burton of Ohio—former chair of the House Rivers and Harbors Committee and a founding member of the NRHC—endorsed a request in the House to investigate the NRHC as "a great lobbying scheme." Burton characterized a flood control and navigation project on the Sacramento River as aid for private land reclamation. Sen. Joseph Ransdell, NRHC president, responded to such criticism by saying that the NRHC is not a lobby as it does nothing concealed or underhanded. Ransdell defied anyone to prove that southerners controlled the organization.[126]

A National Network Forms

Expansion of the river lobby's network of supporters defused criticism of the lobby's sectional imbalance. Flood control activists marshaled existing networks of regional shipping lobbies. They also used decades-old economic and personal connections among planters, New Orleans business leaders, and northerners invested in southern farmlands to create interlocks among regional flood control organizations.[127] Business interruptions due to floods, press coverage of the 1912 flood, and activist reinterpretations of the floods as a national, not local, concern boosted the creative cooperation of river organizations.

The two most influential activist organizations of the time adopted different strategies. The NRHC lobbied most intently for flood control improvements on the lower Mississippi River, but as it had achieved a reputation as a national organization, it also promoted a nationwide flood control program.[128] The MRLA focused on demonstrating that the flooding of the lower Mississippi River was a national problem requiring government aid.[129]

Other organizations supported demands for federal aid. The secretary of the National Drainage Association, Major J. A. Dapray, urged attendees of a NRHC convention to support government drainage of land that was "completely inundated, marshy, swampy wet-land, breathing pestilence [malaria] and adding to the sickness of this country."[130] Shipping activist organizations associated with the NRHC became peripherally involved in flood control activism.[131] Over the years, the Mississippi Valley Flood Control Association and the Board of Mississippi Levee Commissioners had promoted flood control aid through memorials and personal contacts with members of Congress from the valley.[132] The New York Board of Trade and Transportation issued a pamphlet in response to the 1912 floods called "Mississippi Valley Flood Relief: A National Obligation."[133]

Activists also coordinated their lobbying for flood control. In 1913, the NDC organized a meeting of thirty-five organizations from fifteen cities, the Mississippi Valley Improvement Association, the Southern Settlement and Development Association, levee districts, and commercial organizations.[134] The MRLA published a speech by the president of the federal government's Mississippi River Commission about flood control needs on the Mississippi River.[135] The MRLA solicited support from the Southern Merchants Convention, the National Hardwood Lumber Association, the National Association of Box Manufacturers, the Grain Dealers National Association, the National Association of Credit Men, the Southern Commercial Congress, and the American Bankers' Association.[136] The MRLA and the National Rivers and Harbors Congress met in a joint convention in Washington, D.C., in 1914 to protest cuts in that fall's rivers and harbors bill, featuring an opening speech by Secretary of State William Jennings Bryan in favor of federal flood control funding. The meeting prompted an additional river appropriation bill in the spring. Even though the flood control cause had been advanced by the navigation lobby, New Orleans machine boss Martin Berhman and other flood control activists began plans to separate flood control funding from rivers and harbors bills to avoid complaints about pork barrel spending.[137]

In 1916, the lower Mississippi River flooded again. The San Diego River rose high enough to isolate the city of San Diego for a month. The New Orleans Association of Commerce urged Congress to pass a joint resolution stating that flood control on the Mississippi River was a national problem and an obligation of the federal government.[138] House leaders had closed the levee committee in 1911, during a period when the levees had held back the yearly floods. By late 1915, House representatives Charles Curry from the Sacramento Valley and Benjamin G. Humphreys, a long-time activist from the Mississippi Valley, had already persuaded Speaker Champ Clark to create the House Flood Control Committee. Despite protests from Rivers and Harbors Committee members, the House approved the committee while the river flooded in 1916. Humphreys became chair.[139]

Competing Plans

Advocates of central planning and multiple purpose development continued to submit proposals in competition with those made by advocates of the Corps of Engineers. President Wilson had his secretaries of agriculture, commerce, and the interior study flooding on the Mississippi River. In 1916 they recommended a comprehensive development plan financed by United States government bonds.[140] Senators Newlands and Broussard in 1917 reintroduced their bill to create an Inland Waterways Commission to coordinate comprehensive river development.[141]

Some activists and members of Congress hedged their bets by supporting more than one proposal. The NDC supported the Newlands and Broussard bill. In consultation with the secretary of the interior, an NDC committee also wrote an alternative bill authorizing the interior secretary to oversee flood control and drainage works, introduced by House Speaker Champ Clark (Democrat, Missouri) and Sen. John

Sharp Williams (Democrat, Mississippi).[142] On the other hand, Williams—a planter in the Yazoo-Mississippi Delta—supported the efforts of fellow planters John M. Parker and LeRoy Percy to gain Corps flood control aid. Clark had spoken in support of the NRHC and of the Corps' program and had created the House Flood Control Committee.[143]

In opposition to comprehensive planning, NRHC President Senator Ransdell and NRHC member Rep. Benjamin G. Humphreys of Mississippi co-sponsored a bill to grant the Corps the full legal authority to build levees and protective revetments on the Mississippi and Sacramento rivers. This prompted heated debates in pamphlets and in congressional hearings pitting comprehensive, multipurpose resource planning versus single-purpose water development.[144] Humphreys held extensive hearings in the Flood Control Committee, with testimony from activists including LeRoy Percy and A. S. Caldwell of the MRLA; Walter Sillers, president of the Mississippi Levee District in the Yazoo basin; and New Orleans mayor Martin Berhman of the NRHC. To attract western support for a potential flood control bill, committee members heard testimony about the Sacramento River from Senator Newlands and irrigation advocate George Maxwell. In 1916, the committee issued the Ransdell-Humphreys bill with $5 million for the Sacramento River and the House passed the bill.[145]

Some members of Congress attacked this bill as a "stupendous loot of the Federal Treasury" and an attempt by levee districts to get taxpayers to pay for reclaiming private land.[146] North Carolina Rep. John H. Small, later president of the NRHC, asked fellow legislators to acknowledge funds spent on Mississippi River levees for navigation improvements had been intended and used to build flood control levees. This had been acknowledged during other hearings in recent years.[147] Nonetheless, appropriations in the rivers and harbors bill of 1916 for Mississippi River levees, as in most previous years, were made "to improve navigation."[148] Ongoing maneuvering between Senator Newlands and flood control advocates kept the Ransdell-Humphreys bill tied up during the next year, while hostilities between the United States and Germany came to dominate Congress' work.[149]

Efforts to add irrigation funding to the Ransdell-Humphreys flood control bill were blocked when it was reintroduced in 1917. This turned some legislators from the arid West against the bill. Sen. George Norris of Nebraska complained that Congress had offered little support to landowners building irrigation works in arid regions. Norris noted that Congress paid twice as much for federal levees on the Mississippi River as local beneficiaries paid.[150]

The MRLA and legislators from the Sacramento and Mississippi river areas lobbied actively in 1917 for levee aid justified for flood control. Planter LeRoy Percy of the MRLA, who had served in the Senate from 1910 to 1911, spent weeks in Washington, D.C. Percy led a group of MRLA activists and testified before the House Committee on Rivers and Harbors about the superiority of the Ransdell-Humphreys plan.[151] Activists circulated copies of Percy's statement to the public.[152]

With the House having again passed the bill, a Senate filibuster by Norris and others opposed to the impending war provided an opening for Ransdell to press for a special session to consider the flood control bill. Democratic leaders apparently wished to avoid risking southern support for the war and approved the session. Newlands allowed the bill to pass. President Wilson appeared ready to veto the bill for failing to include Newlands's commission for comprehensive development. Secretary of War Newton Baker met with valley senators and may have passed their concerns to the president. Wilson also likely wished to sustain support for full presidential war powers. He signed the first official flood control bill on 1 March 1917.[153] According to flood control historian Arthur Frank, the MRLA was the lobby organization most responsible for convincing members of Congress to vote for the bill.[154] In a letter to his son, LeRoy Percy of the MRLA took personal credit for getting the bill passed.[155]

The 1917 Flood Control Act outlined some of the projects still in use on the Sacramento and Mississippi rivers and their tributaries.[156] Details of the Flood Control Act were negotiated in a similar manner as the rivers and harbors bills were generally negotiated, in consultation with local beneficiaries and local and state governments. Congress directed the Corps to build facilities along the Sacramento River for $5.6 million—an amount reduced in committee from the $11 million recommended in Thomas Jackson's California Debris Commission proposal—and on the Mississippi River for $45 million.[157] The Sacramento plan included the Yolo Bypass and Sutter Bypass. Local governments and state governments were required to pay one-third of the costs for constructing the levees. They were also required to provide land for the projects and to pay for rights-of-way on privately owned properties.

Rights-of-way would be expensive in sections where engineers wanted to set levees far back from the river, onto crop lands.[158]

The Board of Engineers would review project policy and decide how to distribute levee building costs between local and state governments. States were to pay for maintenance. In a nod to those favoring multi-purpose planning, the act also required that the Corps assess the feasibility of beginning hydropower, irrigation, and soil conservation projects on the Sacramento and Mississippi rivers, as well as navigation and flood control. The federal Mississippi River Commission would oversee project work and hire contractors to complete the work. Military engineers from the Corps would directly supervise all work crews. This act decisively established federal government responsibility to protect lands adjacent to navigable rivers, and it further institutionalized relations between the federal government, contractors, and state and local governments.

Conclusion

Republicans reshaped national policy to promote business and to make government more "efficient." River activists successfully associated their own demands with these goals to win the first formal flood control act in 1917, even though it hardly represented Progressive principles of efficient management by experts. Improved opportunities for travel and communication had allowed favor seekers to build stronger links to local organizations throughout the country, enabling them to back their rhetoric about the national interest with a nationwide organizational base. By organizing a national coalition to support aid to the two valleys, river advocates reduced the appearance of self-seeking. This marked the maturing of flood control activism into a lobbying operation better able to respond to the Progressive political rhetoric of the day.

In his legislative history of the 1917 Flood Control Act, Arthur Frank identifies many government agencies, private organizations, and social trends as causes but provides no framework for explaining their relative importance or how these conditions developed historically.[159] As the previous chapters show, the 1917 Flood Control Act was made possible by a long history of legal and political precedents. These events expanded central government intervention into land development and elaborated ties between government agencies and economic actors that yielded the articulated set of institutions responsible for implementing the flood control program.

In the years after they won the 1917 Flood Control Act, advocates for flood control aid worked mainly to sustain appropriations and to institutionalize the flood control program within Congress and the Corps of Engineers, although they retained an activist sensibility. Their work with legislators and army engineers, and their contact with supporting groups throughout the country, enabled advocates to mobilize support for more extensive river plans following a devastating flood in 1927.

chapter 9

The Fully Designed River

It is perhaps the greatest engineering feat the world has ever known.—Frank Reid, Chair of the House Flood Control Committee, on the 1928 Flood Control Act, quoted in Barry, *Rising Tide*, 406

Redesigning Rivers

For a time after the 1917 Flood Control Act, activists who were concerned mainly with the Mississippi and Sacramento rivers considered their main work done. Funding became their key concern. After the great Mississippi flood of 1927, activists reemerged to demand that the federal government redesign the Sacramento River and the lower Mississippi River. For activists critical of the levees-only policy, the 1927 flood disaster confirmed that Congress and the Corps had failed to fulfill their essential duties of providing sufficient appropriations and designing effective river controls. Advocates' demands took on moral fervor, emphasizing the government's duty to protect lives as well as property.

Campaigners and sympathetic politicians simply presumed that the government would respond to the 1927 flood. They communicated this presumption so effectively that public debates concerned how the government should respond, not whether it should respond, despite President Calvin Coolidge's resistance. Hearings and news accounts centered on which engineering approach would best control each river. Despite their differences, each of the plans that were seriously considered aimed to redesign the rivers. This is what Congress directed the Corps to do in 1928.

Debates about the 1927 flood turned the local and regional problem of flooding into a national issue. This had several consequences. The Corps of Engineers changed. Rhetoric about the national interest had long appealed to top army engineers, and the 1927 flood hurt the army engineers' pride. The 1928 Flood Control Act marked the point when the Corps of Engineers embraced flood control as a duty fully compatible with its navigation improvement work. Presidents and members of Congress would now likely be expected to respond to all large flood disasters. And the 1927 flood made it possible for legislators to require the federal government to share flood control costs. All of this became possible because activists had been working for three decades to make the flood control campaign into a civic venture. Since founding their professional lobby organizations after the turn of the century, activists had spoken less about the need for regional growth in the Sacramento and Mississippi valleys and more about reducing hazards and responding to public disasters. The 1928 Flood Control Act made these redefined goals into national policy. If flood protection was in the national interest, advocates of flood control were no longer serving their own interests when they demanded greater flood protection for specific rivers.

Interest Groups Rise

During the early twentieth century, ideological distinctions between the parties continued to fade. Progressive Movement reformers within the Republican Party contributed to the general decline in party influence by decreasing party opportunities for patronage, instituting direct primary elections, and curtailing the power of the House Speaker. As a result of the latter reform, standing congressional committees and subcommittees could act with greater discretion. Legislators consulted directly with the Department of Agriculture, the Bureau of Reclamation, the Corps of Engineers, and the interest groups that benefited from these agencies' programs.[1] Business interest groups saw new opportunities for influencing congressional votes. These conditions encouraged a focus on issues that appeared to be primarily technical.

Congressional reforms strengthening the role of committees may have also made it more difficult for would-be planners to create programs for comprehensive resource planning. Interest groups that backed the Corps of Engineers certainly resisted proposals for comprehensive planning in their interactions with members of Congress. Senator Francis Newlands won congressional approval of a Waterways

Commission by threatening to block the 1917 flood control bill. With Newlands and other supporters of multipurpose planning divided over policy options, however, the permanent commission was never created.[2]

Some elements of multipurpose development were adopted. Gifford Pinchot and other advocates of public power finally got the Water Power Act through Congress in 1920, establishing federal regulation of all potential water power sites. The multipurpose Hoover dam was begun in 1928.[3] But the new Federal Power Commission had little real authority over hydropower developers. The act had failed to specify how federal resources agencies were to coordinate power development with other federal water projects. According to Donald Swain, the Water Power Act was a weak substitute for Newlands's program of comprehensive waterway planning.[4] Congress continued to assign new water projects to existing agencies, including the Corps of Engineers.

Lobbyists Try to Engineer the Engineers

In the view of flood control advocates in the 1920s, funding for flood control was inadequate, and engineering choices made by the Corps were questionable. The Mississippi River Levee Association disbanded after Congress passed the 1917 Flood Control Act, considering their goal won. The National Rivers and Harbors Congress (NRHC) suspended its meetings from 1917 until 1922, during World War I and a depression.[5] War and depression also constrained government spending. Only $17 million of the $30 million appropriated for the Mississippi in 1917 went toward levee building, due to wartime inflation and the cost of emergency work during floods in 1920 and 1922.[6] Activists began organizing once again to ensure funding and to influence project design.

The process of building unified flood control systems stimulated different political responses in the Sacramento and Mississippi valleys. On the lower Mississippi River, people began to promote designs they felt would best benefit their specific upstream or downstream location. The shortage of funding to build levees or to test alternatives, however, made it difficult for advocates to decide among engineering options for the Mississippi. Engineering methods for the Sacramento River had become uncontroversial, although funding did remain a problem.

Many observers had come to consider outlets and reservoirs to be viable for the Sacramento River system.[7] The chief activist organization, the River Improvement and Drainage Association, had called for a levees-only policy at the turn of the century, but it became evident that natural outlets from the Sacramento River had historically formed to carry off excess water during moderate floods. There was little settlement pressure in the areas of these natural outlets. Levee districts and California state engineers therefore had little interest in closing the outlets. Landowners in some of the proposed bypass areas did resist government efforts to build enhanced bypasses, but the California Supreme Court ruled to allow the bypasses. State and local agencies completed the Sutter and Yolo bypasses in the 1920s.[8]

By the 1920s, effective debris dams had been built in the Sierra Nevada range, and powerful dredgers had spent years removing silt from the Sacramento and its tributaries. Few thought it economically or politically feasible to restart hydraulic mining. Valley interest groups made sure by soundly defeating occasional proposals made for additional debris dams to capture newly generated mining debris.[9]

About 700,000 acres were reclaimed in California's Great Central Valley by 1918, with protection most complete in the American River basin.[10] Large-scale land companies had formed viable reclamation districts in many areas. Farmers on the best lands had established high-yield orchards and row crops. Farm size actually decreased by the 1920s in some intensively farmed areas. Intensive operations required access to cheap labor. The workforce included Mexicans, East Indians, and other immigrants from overseas. In 1934, the Dust Bowl provided a surplus of white migrants from Oklahoma and other states, displacing other workers.[11] River shipping increased, with 90 percent of the freight from Sacramento to San Francisco passing down the Sacramento River.[12] Under these conditions, lobbying by northern Californians continued to focus on increasing federal appropriations and was done largely by congressional delegates and by those who joined national organizations.

In contrast, activists from the Mississippi Valley began to promote a variety of specific engineering approaches and designs, as some of the economic and political consequences of various project designs became evident. Members of the well-intentioned public delighted in submitting engineering suggestions. These included fanciful plans to boil off the flood waters or to build a canal of unprecedented length from the Ohio River to the Gulf.[13] Among the methods considered more plausible by engineers, several attracted widespread support.

The federal government's Mississippi River Commission had adapted the 1861 recommendations of military engineers Humphreys and Abbot, and by the 1920s had come to favor a levees-only policy.[14] According to the commission, concentrating the water into the main channel would force the river to scour out its own channel and increase the channel's carrying capacity.[15] Most mid-river and downstream residents accepted the levees-only approach until the flood plane (river level) increased, around the turn of the century.[16] The 1912 flood broke flood plane records at seventeen of the eighteen river gauges south of Cairo, Illinois, even though its volume was far less than the volume of the 1882 flood. This indicated that the levee-constrained river was failing to scour a deeper bed. Despite debates among commissioners, the Mississippi River Commission maintained its levees-only policy and allowed local levee districts to close Cypress Creek below the Arkansas River in 1921.[17]

Civilian engineers, activists, and officials increasingly critiqued the Mississippi River Commission's approach. Gifford Pinchot, Teddy Roosevelt, New York Gov. Franklin D. Roosevelt, and many residents of upstream areas supported watershed reforestation and headwater reservoirs. The independent engineer James Eads had argued in the mid–nineteenth century that cutoffs forcing the meandering river into a straighter channel would improve flood control, but this view attracted few advocates.[18] A committee of the American Society of Civil Engineers approved of levees-only for the flat, alluvial portions of the lower Mississippi River, but the report failed to mollify civilian critics of the Mississippi River Commission. The New York Board of Trade and Transportation distributed a mailer in 1914 rejecting multipurpose development as too complex, rejecting the levees-only approach, and promoting spillways to divert flood-stage waters around New Orleans and other important sites.[19] The president of the federal Mississippi River Commission, army engineer C. M. Townsend, went on record supporting a multipurpose development, although his proposals would have meant little change in the commission's flood control policy for downstream alluvial lands. Townsend proposed reforestation efforts to help conserve soil; reservoirs to regulate river levels for shipping, to generate power, and to provide water supplies; and levees for flood protection.[20]

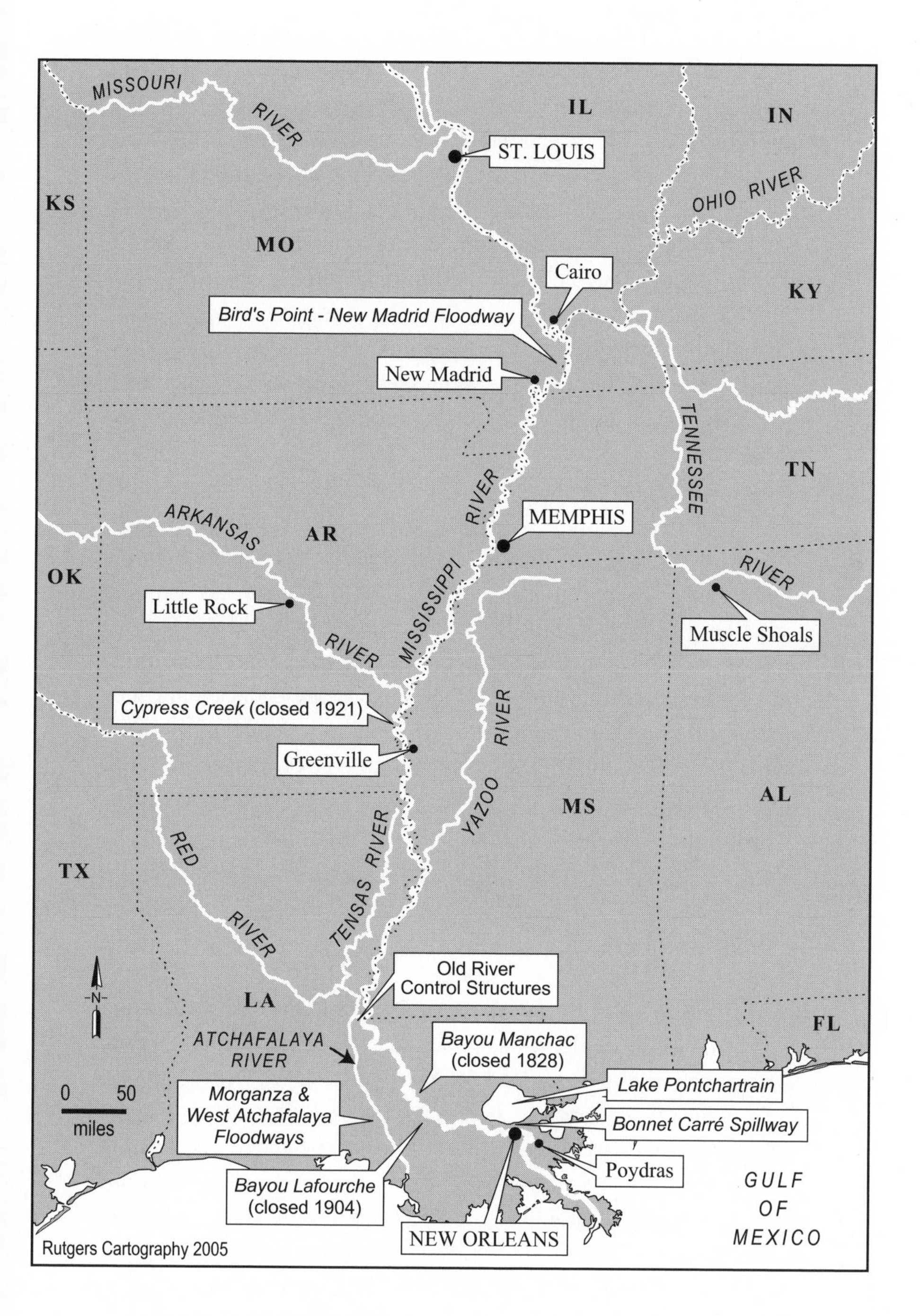

MAP 3. Lower Mississippi Valley

A system of levees relieved by strategically placed artificial outlets emerged as the major alternative to the levees-only policy. Many outside engineers, a few Corps engineers, and growing numbers of downstream residents began to advocate various combinations of outlets and spillways. Outlets would constantly drain water from the main channel, and spillways would divert water only during heavy floods.[21] Local landowners and government officials had, however, already closed off many lateral bayous that emptied to the Gulf. Because river outlets took up potentially productive lands, Corps field officers faced strong local opposition whenever they proposed keeping an outlet open or opening a new outlet.[22] Even so, the outlet theory received new attention as the river level rose.

Apart from posing a challenge to the Corps' engineering expertise, each engineering solution seemed to favor a specific constituency. In the early years, levee building had pitted neighbor against neighbor. As the levee lines became more complete, upstream residents achieved greater local protection, but downstream residents could suffer from the levee-only system, because it caused the river to carry a much greater volume of water. Given that volume, a break in the tall, modern levees could cause tremendous local devastation.[23] On the other hand, the building of an outlet or spillway in a downstream location would require the government to appropriate private lands. Perceptions about engineering choices began to affect political alignments.

Lobbyists Diversify

In the spring of 1922, the lower Mississippi River flooded again, quite severely in places. The river was so high that it pushed tributary waters back up their channels to flood six Yazoo-Mississippi Delta counties. While managing emergency levee work in the Delta, activist planter LeRoy Percy contacted bankers around the country for loans. He formed the nine-member West Mississippi Flood Committee to lobby Congress, comprised of a leading planter from each Delta county. Telegrams poured in from officials and business people along the river. Congress soon appropriated $1 million to aid emergency levee rebuilding.[24]

Coming so soon after the Cypress Creek outlet in Arkansas had been closed, the flood newly provoked critics of the levees-only policy. At a sharp turn in the river at the French Quarter of New Orleans, water from the 1922 flood had seeped under the levee and burst through the cobblestones like a small volcano, alarming residents of New Orleans. When a levee break at Poydras, twelve miles south of New Orleans, dramatically lowered the river at New Orleans, city leaders began planning to make that outlet permanent.[25]

New Orleans newspaper publisher James Thomson championed a civilian engineer's assertion that the Poydras crevasse had saved New Orleans. Thomson called a mass meeting to discuss rebuilding after the 1922 flood. Leaders of unions, New Orleans banks, the Cotton Exchange, the Board of Trade, and the Association of Commerce at the meeting formed the Safe River Committee of 100. They called for new outlets to the swamps and to the Gulf. Under the guidance of executive director Walter Parker, a board member of the Association of Commerce, members sent literature and telegrams to elected and appointed federal officials, testified before committees of Congress, wrote editorials, and presented technical reports about outlets at meetings of civil engineers. New Orleans bank presidents and trade association leaders who served on the committee also worked behind the scenes to lobby Congress and President Warren Harding.[26] The Board of the Orleans Levee District asserted that the Louisiana state government would pay to build a spillway below New Orleans, if the federal government would approve it.[27] The Safe River Committee also lobbied in 1924 and 1925 for Sen. Francis Newlands's bill to create a commission to take over river development coordination from the Corps. The Chief Engineer of the Corps, Harry Taylor, naturally opposed this bill.[28] The Safe River Committee of 100 soon became the most influential organization in the country advocating outlets.[29]

In 1922, Chief Engineer Gen. Lansing Beach responded with irritation to his critics. He asserted that New Orleans leaders sought the spillway only because they wished to avoid rebuilding the city's ports up to the new standards required by the Mississippi River Commission. Then, after another round of criticism, Beach told New Orleans business leaders at a 1922 spillway meeting: "If it were my property, I would rather blow a hole in a levee, if conditions became serious, and let the water take care of itself, rather than [pay to build the spillway] and pay $250,000 a year continually in interest charges [for bonds] and the additional cost of maintenance."[30] Nonetheless, the official policies of the Corps-affiliated Mississippi River Commission continued to favor levees. Levee breaks had lowered the flood plane during the 1922 flood, but those levees had been below the Mississippi River Commission's quality standard. Army engineers judged that a completed system of standard levees would solve the flooding problems. They began plans to close the last and largest outlet, the Atchafalaya River.[31]

A representative from New Orleans introduced a bill for James Thomson in 1922 to study reservoirs, cutoffs, and spillways. The bill generated four years of heated debate, and it divided activists. LeRoy Percy's engineering advisor, Mississippi River Commission member Charles West, opposed spillways near New Orleans because they would likely harm the Yazoo-Mississippi Delta and other upstream areas. Percy worked to convince levee district managers and business leaders to support the levees-only plan, but after 1922, he could rally few of his old allies to support that technique.[32]

When levee districts on the lower Mississippi organized yet another association after the 1922 flood, they proposed a variety of flood control techniques. Knowing that the 1917 Flood Control Act was about to expire, members of levee boards created the Mississippi River Flood Control Association. The association assessed dues from levee boards according to their ability to pay. Using methods similar to those of the predecessor levee association, the Mississippi River Flood Control Association requested that business and industrial firms in the region encourage their employees to spread the association's propaganda calling for renewed aid.[33]

Other organizations also responded to the 1922 flood. The Mississippi Valley Association had united in 1919 under the uninspired slogan, "All for one and one for all." The past president of the National Drainage Congress, Frank B. Knight, had traveled to New Orleans with a delegation from the Chicago Association of Commerce to boost commerce in the Mississippi Valley, and they formed this new association. The Mississippi Valley Association called for the Corps to complete all unfinished navigation projects and for Congress to create new projects for navigation, land reclamation, and flood control.[34] George H. Maxwell, head of the National Reclamation Association, warned that the lower Mississippi Valley would suffer a "cataclysm" unless comprehensive river planning were undertaken. The U.S. Chamber of Commerce sent a resolution to Congress calling for the control of Mississippi River floods.[35] Sacramento Valley activists meanwhile renewed their call for increased federal payments for the Sacramento River. Under the 1917 Flood Control Act, the federal government paid a smaller proportion of the costs for the Sacramento River than it did for the Mississippi River, in part because the Sacramento did not cross state lines.[36]

Most significantly, members of the influential NRHC in 1925 adopted their first extensive set of resolutions favoring a permanent federal flood control program. They formally enlarged the NRHC's objectives to include flood control and other forms of water development.[37] The NRHC, U.S. Chamber of Commerce, Mississippi Valley Association, Intracoastal Canal Association, and other organizations lobbied for a nationwide program of river surveys for multipurpose development. Organizations speaking to the NRHC in favor of flood control included the National League of Women Voters, the Ohio Valley Improvement Association, and the Mississippi Valley Association.[38] It was estimated that states, levee districts, and other local agencies had spent $165 million on flood control between 1881 and 1925, while Congress had spent $65 million.[39] A representative from Arkansas called these local payments "tribute" that property owners elsewhere would not have tolerated.[40] The special counsel to the Orleans Levee District remarked, "The port of New Orleans asks Uncle Sam for nothing except a square deal."[41]

"Psychopathic Enthusiasts"

Critics of the Corps and of the NRHC argued that even little-traveled streams received aid for navigation improvements.[42] Although the U.S. Chamber of Commerce had long lobbied for river aid, its leaders sent a letter to members in 1923 calling the rivers and harbors bill "the famous 'pork bill' of each session" and suggesting the Corps was corrupt because it made project recommendations to Congress.[43] In 1926, an editorial in the *Engineering News-Record* said that most members of the rivers and harbors organizations were "psychopathic enthusiasts."[44]

President Calvin Coolidge praised river navigation and flood control development in his 1924 State of the Union message. Mindful of the critics, Coolidge also said he would refuse to sign a bill that included projects he thought unnecessary.[45] Sen. Joseph E. Ransdell, former

NRHC president, reported in 1925 that members of Coolidge's cabinet had attended meetings of the Missouri River Improvement Conference, the Mississippi Valley Association, and the American Farm Congress to promote water development. He asserted that public sentiment for waterway improvements was the highest in twenty years.[46]

Congress passed a bill in 1925 backed by advocates of multiple use development. It ordered the Corps and the Federal Power Commission to jointly study navigation, flood control, irrigation, and power development on all of the country's navigable waterways where power development appeared feasible.[47] The data published in these "308" reports, named after the House document number,[48] would guide debates over New Deal waterway development for years. But from 1924 through 1933, Congress approved few new projects and directed its annual navigation appropriations toward completing existing projects.[49]

Residents and leaders of New Orleans—who lived below sea level—had gotten the state of Louisiana to build a spillway at Pointe a la Hache south of the city in 1926, and they continued to seek additional spillways. On the other hand, levee engineers and residents upstream had begun to feel confident in their levees. Yazoo-Mississippi Delta residents created over one hundred drainage organizations to prepare cotton lands behind the protective federal levees. High demand for cotton during World War I led to active sales of leveed lands and increased tax revenues to the levee districts. The Mississippi River Commission in 1926 predicted that the completion of the federal government's levee building project on the Mississippi River was approaching.[50] Even so, in 1926, after four years of bitter wrangling between proponents and opponents of spillways, Congress created a spillway board. Its members would make an investigative field trip to New Orleans, scheduled for the spring of 1927.[51]

A spirited convention of the NRHC in December 1926 reflected the high expectations held by advocates of the Corps of Engineers. The chief of engineers, Maj. Gen. Edgar Jadwin, commented on the Mississippi and Sacramento river projects. He also noted that Congress was then considering flood control problems on certain other streams. Later, descendants of Ulysses S. Grant and of Robert E. Lee were presented, accompanied by a band playing "Yankee Doodle," followed by "Dixie." General John J. Pershing led the band in the "Star Spangled Banner," and Sen. J. Thomas Heflin of Alabama told "Stories from the

South," which turned on the alternately wily and foolish behavior of African American characters.[52]

The Great Mississippi Flood of 1927

The 1927 flood of the Mississippi River crested ten distinct times at Cairo, Illinois, made at least seventeen large crevasses along the lower river, and was sustained during one period for 153 consecutive days. Storms began in August 1926 in the Midwest and South, leaving twenty-one dead by October. Gauge readings of the river plane on the Ohio, Missouri, and Mississippi rivers were the highest three-month averages ever measured at those sites. Rain began again in mid-December producing heavy flooding. By February, localized floods in the lower valley killed another thirty-two and tornadoes killed forty-five. The remaining levees held, but the river was carrying more water than ever. It had only two remaining outlets, the river's main channel to the Gulf, and the Atchafalaya River, which branched from the Mississippi River and led to the Gulf through the swamps.[53]

Crews of African American sharecroppers worked at many places along the lower river—sometimes at gunpoint—putting up supplies, fixing weak spots on levees, and sandbagging levee tops. Armed white men ran patrols to prevent people from the other side of the river from cutting or blowing up a levee, which was attempted in several places. In the Yazoo-Mississippi Delta, the Mississippi National Guard assisted. Delta planter LeRoy Percy had the Mississippi governor send convicts to do physical labor. Refugees began to gather, totaling 50,000 along tributaries and the main channel by April. They lived in box cars, army camps, and elsewhere.[54] Engineers pulled a train with twenty-one coal cars onto a bridge in Little Rock to steady it. Just when vibration from the Arkansas River set the coal on fire, the bridge and train collapsed into the river. New Orleans received twenty-five percent of its yearly average rainfall on a single day in April, the most ever recorded. The river carried between two-and-a-half to three million cubic feet of water per second past the Yazoo-Mississippi Delta in April, nearly three times the amount of the 1993 floods.[55]

The day after a major crevasse in the Yazoo-Mississippi Delta, as New Orleans residents approached panic, President Coolidge finally responded to valley governors' pleas. Coolidge designated Commerce Secretary Herbert Hoover chair of a cabinet committee to organize rescue and relief efforts.[56] Coolidge ordered the Mississippi River Commission in May to outline a full plan for flood control on the lower Mississippi. Chief Engineer Jadwin directed the commission to study reforestation, reservoirs, and spillways as well as levees.[57]

No recorded river flood in the United States has been close to the size of the 1927 flood. It remained unsurpassed in damage costs until the 1993 flood on tributaries of the Mississippi River.[58] So much water passed into the lower river valley in 1927 that lands miles beyond the usual flood plain were inundated.[59] Federal officials estimated that 700,000 persons were driven from their homes. Another 330,000 were rescued from the tops of trees, levees, and houses by boats run by government agents, fishers, and bootleggers. The U.S. Weather Bureau estimated direct property losses to be $360 million. The Mississippi River Flood Control Association estimated direct losses at $230 million and indirect losses at $200 million.[60] The river destroyed every bridge for a thousand miles. It broke levees in thirteen places south of Cairo, Illinois, and flooded about twenty thousand square miles, affecting every major basin of the lower valley. Officials estimated that the floods killed 250 (the validity of this estimate is unclear), with African American tenant farmers suffering the greatest loss of life.[61]

Newspapers around the country published many articles and photos of the devastation. Thousands of editorials were written about federal aid, mostly in support of aid. The chairman of the American Picture Show Association urged Congress to pay the entire cost for the projects. He admitted that his association was spreading propaganda in favor of aid by showing flood scenes at theaters. Many radio stations gave hourly reports on the flood at its worst. Up to forty radio stations at a time linked together to broadcast speeches about the flood, including several speeches by Secretary Hoover.[62]

Hoover Rises to Fame

This was Herbert Hoover's equivalent of San Juan Hill, a sixty-day opportunity for constant press attention as he set up refugee camps and relief operations.[63] Hoover had appeared before the NRHC in previous years to advocate surveys and improvements of the Mississippi River. He became widely known for these positions only during the 1927 flood.[64] Before the flood, Hoover had not been considered a notable candidate for president. After the flood, Hoover was the standout candidate.[65]

The federal government had seldom been involved in giving disaster aid. Hoover's approach to flood relief reflected the priority that the Coolidge administration set for promoting business growth aid, although Coolidge came to disagree with Hoover's growing enthusiasm to build flood control works.[66] Hoover—who had managed war relief aid for Belgium—arrived in April and began to plan for relief efforts. Following the practice of local planters, the Red Cross and National Guard units under Hoover's control at several of the refugee camps kept African American sharecroppers in the camps by force until needed for work on levees or on flooded fields. Hoover's attempt to encourage private credit for planting largely failed. Few sought credit, 70 percent of the crops failed, pellagra spread widely in the refugee camps, and distribution of relief supplies through planters resulted in African American farmers, tenants, and sharecroppers receiving very little aid.[67]

A Renewed Lobbying Campaign

Politicians, editors, and activists around the country pressed Coolidge to call a special session of Congress for more substantial relief aid. Hoover met with LeRoy Percy and asked him to reverse his public support for this session, implying that if Hoover were elected president, he would not promise to produce a flood control bill. Percy, New Orleans banker James P. Butler Jr., and other valley leaders whom Hoover had contacted then announced that they thought the session was unnecessary. Calls for the session died down.[68]

In the meantime, using their connections to the statehouse, the Corps, and Congress, New Orleans elites got the decision that they sought. Bankers James Butler, Rudolph Hecht, and J. Blanc Monroe, and the founder of the Safe River Committee of 100, James Thomson, joined with the past and current presidents of the NRHC, Sen. Joseph Ransdell and Rep. Riley Wilson (Democrat, Louisiana) to apply pressure to the Corps.[69] Crevasses elsewhere may have already lessened the threat to New Orleans. Nonetheless, these men got permission from the Corps to blow up a levee below New Orleans on June 12. The crevasse flooded a poor and lawless rural parish that relied on fur trapping and bootlegging.[70]

LeRoy Percy contacted Lewis Pierson, president of the New Orleans Chamber of Commerce, and New Orleans bankers and lawyers to form a national campaign.[71] Percy, banker Rudolph Hecht, and publisher James Thomson were on the executive committee for the Chicago Flood Control Conference in June, organized by Chicago Mayor William Thompson. Although Dwight Davis, President Coolidge's secretary of war, discouraged legislators from attending, the Chicago meeting attracted two thousand delegates in mid-flood for the purpose of demanding a federal program for the Mississippi River. As a member of the convention's executive committee, and as a member of the U.S. Chamber of Commerce's flood control planning committee, Percy began to travel extensively. He met with President Coolidge, Chief Engineer Jadwin, and other officials around the country. At his home, Percy hosted Vice President Charles Dawes, and he led a delegation of bankers and manufacturers on a tour of flooded areas.[72]

National organizations sent resolutions to Congress or distributed educational material to the public in response to the flood. These included the Chicago Association of Commerce, the National Association of Manufacturers, the Infant Children's and Junior Wear League, the American Bankers Association, and the Investment Bankers Association of America. Some groups sent special trains carrying delegations to the capital.[73]

Aid proponents argued that levee boards along the Mississippi had issued bonds far beyond the total valuation of their districts and that the federal government must now finance all federal plans.[74] The Mississippi River Flood Control Association distributed a report on the financial losses and physical damage that the flood caused. They highlighted losses in industries such as railways that would affect investors from around the country.[75]

The Chicago Association of Commerce asserted that because the 1917 Flood Control Act had led private interests to invest millions in swamplands along the river, it had established a further obligation for the federal government. People from around the country who had interests in southern real estate, farming, crop processing, or lumber operations or held bonds issued by southern levee boards or drainage districts raised public concern by speaking in their home communities about the flooding problems. National associations of real estate boards, timber cutters, manufacturers, banks and other industry groups with investments in the Yazoo-Mississippi Delta also sent resolutions to Congress. Banks from cities in the Delta paid for their employees to lobby Congress in Washington. Reports on the economic effects of Delta flooding from sources such as Bradstreet and the Guaranty Trust Company of New York provided advocates ready data for propaganda.[76]

The 1928 Flood Control Act

Historian Matthew Pearcy argues that the 1928 plan became possible mainly because of the leadership of politicians from the upper Mississippi region who promoted an array of techniques to improve flood control on the Mississippi River system. Politicians and activists from the lower river who had favored or at least accepted the levees-only policy of the Mississippi River Commission had lost credibility with the 1927 flood. President Coolidge's resistance to aiding the valley provoked Republican Frank Reid of Illinois to push his own vision for the river within the Republican Party, which depended on Illinois, and within Congress.[77] As we have seen, Illinois politicians had long promoted ambitious water schemes. Some of the conventions and organizations they had set up over the previous century had promoted river and rail aid exclusively for the northern Mississippi valley, but at other times, they joined activists from the South in demanding improvements for the Mississippi River. Pearcy notes that interest in the river was high in Illinois, because Congress had just approved the long-sought Lakes-to-the-Gulf Deep Waterways project and because the 1927 flood had devastated Illinois farmland.

The House Committee on Flood Control began hearings in November, soon after rescue work had ended.[78] Chicago Mayor Bill Thompson attracted great publicity with his delegation to Washington and his testimony about his firsthand experiences of the flood. Hoover testified about the national economic consequences of the flood. Rep. Philip Swing and Flood Control Committee chair Frank Reid of Illinois attacked the Corps' stubborn protection of its levees-only policy against countervailing evidence. Gifford Pinchot called the Corps' levee policy the "most colossal blunder in engineering history." Forestry officials outlined plans for upstream reforestation as a supplement to flood control engineering. Witnesses representing New Orleans called for stronger levees and new spillways. Witnesses from areas near tributary streams called for reservoirs. Chief Engineer Jadwin defended the levees-only policy.[79] The governor designated Percy as Mississippi's official representative. Percy, James Butler, and James Thomson stayed in Washington, D.C., for long periods to lobby officials during the hearings. Frank Reid sponsored the bill they favored.[80]

The Mississippi River Commission and the chief engineer worked on proposals for the flood control system that fall. But when the commission presented its report to Chief Engineer Jadwin, he sought to control project revisions by proposing that the army unite behind his own proposal, which had Coolidge's implied support as an option that was fiscally restrained, at least relative to the alternatives. The Mississippi River Commission proposed permanent outlets at several points along the river. Jadwin proposed fuse-plug levees instead, designed to burst and divert excess water only during periods of very high water. The plan that Jadwin submitted to the House Flood Control Committee included a floodway near New Madrid (to protect Cairo), the Boeuf Floodway (to replicate the Cypress Creek outlet), a floodway alongside the Atchafalaya River (to lower the flood plane above New Orleans), and the Bonnet Carré Spillway into Lake Pontchartrain (to protect New Orleans). Historians Charles Camillo and Matthew Pearcy conclude that Jadwin's plan advanced not only the fiscal conservatism that he shared with Coolidge but also his desire to gain greater Corps' control over the work of the semi-autonomous Mississippi River Commission.[81]

The chair of the flood control committee of the U.S. Chamber of Commerce, Frederick A. Delano, the uncle of Franklin D. Roosevelt, appeared before the NRHC annual convention. Delano reported that his committee had urged all local chambers to pass resolutions in favor of federal levees on the country's "main trunk-line sewer." Gifford Pinchot delivered a disjointed but passionate speech to the convention. He attacked the Corps for its history of patching together the flood control system and called for a presidential commission to direct a coordinated water development program. Rising to applause after Pinchot's attack, former Chief Engineer W. H. Bixby retorted that the "Army Engineers were the servants of Congress" and had received no mandate for flood control until 1917. Current chief engineer, Edgar Jadwin, talked about the Corps' proposed plan for Mississippi River flood control. The NRHC secretary treasurer reported that the NRHC's "sane and conservative" methods included sending tens of thousands of publications on navigation and flood control to organization members, newspapers, commercial organizations, and libraries. Candidate Hoover complimented the NRHC's work and called the Mississippi River the most urgent of our problems.[82]

The House Flood Control Committee heard over three hundred witnesses by the time hearings ended in February 1928. All but a few of them vigorously critiqued the Jadwin Plan, especially its requirement for continued local contributions to levee costs.[83] Lobbyists focused their efforts on specific plans. The NRHC resolved to support the Jadwin flood control plan.[84] Louisiana Governor Huey Long criticized the Jadwin plan in a letter to Senator Edwin Broussard of Louisiana co-signed by the chief engineers of all levee districts on the lower Mississippi River. Coolidge favored the Jadwin plan, as it was the cheapest.[85] The chair of the House Flood Control Committee, Frank Reid criticized Jadwin, and by extension Coolidge, arguing that Jadwin had illegally produced his own plan. Reid wrote a bill that adapted the Mississippi River Commission plan and that required no local contributions. The committee considered seriously only the Mississippi River Commission plan and the Jadwin plan.[86]

As deliberations went on, Reid talked of preparing a second bill with a nationwide flood control program. Rivers in New England, the plains, and the Far West had also flooded that year.[87] Members of Congress from the greater Mississippi watershed tried to add projects in their own states to the Mississippi River bill. Officials from New Mexico, Oklahoma, North Dakota, Montana, Kansas, and Ohio demanded flood control aid.[88] Conservative representatives from New England asked for no aid for their own region, however, and advocates of Mississippi River aid resisted considering projects for any rivers but the Sacramento. The committee therefore ended consideration of a bill for a national program.[89]

By trading votes with other members of Congress on a variety of issues, Mississippi Valley legislators moved the bill through Congress, adding Sen. Wesley L. Jones of Washington as co-sponsor. Senators in favor of flood control aid had managed to pack the Commerce Committee, and this plus weeks of testimony had turned Jones away from requiring local contributions.[90] Legislators added to the bill a new phase of the Sacramento River project that the federal California Debris Commission had already designed. After wrangling over local cost sharing, a bill favoring the Jadwin plan with a limitation on local costs passed the Senate unanimously. In the House, Reid reluctantly agreed to the Senate version. Coolidge had Jadwin write a critique of the bill and complained to the press that it was a pork barrel bill. The House passed the bill by a two-thirds vote in April, having reduced the cost of building emergency floodways by allowing targeted lands to stay in private hands and compensating owners with a one-time "flowage" payment. Coolidge pressured the House and Senate conference committee to reduce the areas of land subject to compensation in floodways, an element in the bill written in ambiguous language that would lead to later disputes.[91]

Throughout this period, Mississippi Valley activists pressed organizations and businesses outside the valley to lobby Coolidge. Levee boards owed $819 million in bonds. LeRoy Percy, Lewis Pierson of the Chamber of Commerce, James Butler, and other New Orleans bankers told investment bankers that aid would help ensure their repayment. The American Bankers Association passed a resolution supporting the Jones-Reid bill. The Investment Bankers Association, which had already passed a resolution, lobbied Coolidge directly, as did Chicago's Mayor Thompson.[92]

Under pressure from Republican members of Congress who feared the outfall of a veto, Coolidge finally accepted the argument that past local and state payments constituted their share of the levee costs. He signed the Jones-Reid bill in May 1928. Frank Reid called it "the greatest piece of legislation ever enacted by Congress."[93] Not to be topped, Senator Ransdell called it the greatest piece of internal improvement legislation "since the world began."[94] With an authorization of $325 million, it was, at least, the most expensive single bill passed by Congress; the Panama Canal had been authorized for $310 million.[95]

The 1928 Flood Control Act approved the basic plan still in use for the Sacramento River and many of the projects that have been built on the Mississippi River. The Corps, rather than levee districts and state governments, would design and build all levees on the lower Mississippi. Levee districts and state governments on that river would pay no further costs of levee building. Those on the Sacramento River would pay only a small portion of the levee costs, a remarkable benefit at that time for a river that does not cross state lines.[96] Levee districts would continue to pay for rights-of-way. This act set a precedent for the New Deal, by requiring the federal government to actively intervene in local and regional economic development.[97]

The Hoover Administration

Herbert Hoover (Republican, 1929–33) won the Republican nomination a few months after the flood control bill passed, won the presidency, and then came up against the Great Depression.[98] Republicans had promoted direct government aid in the nineteenth century as a means for promoting business growth. By the 1920s, Hoover and many other Republicans tended to favor only indirect government action. This position became especially attractive after the Socialists (in 1908) and the Democrats (in 1928) proposed in their platforms public works programs to employ the unemployed and stimulate the economy.[99]

The liberal conservationism of John Muir, Teddy Roosevelt, and Gifford Pinchot had faded in national politics, displaced by business-minded conservationism.[100] Hoover increased funding for water development and other public works to $2.1 billion. He rejected additional public works spending that would have totaled another $2 billion, calling it a "gigantic pork-barrel."[101]

Congress and Hoover expanded the Sacramento River flood control project in 1929 and authorized flood control work on the Allegheny, Monongahela, and Missouri rivers.[102] Hoover got the Corps into a better position to compete with other agencies for multipurpose projects by having the boundaries of Corps districts remapped to match river drainage areas. He got appropriations for work on the 1928 Mississippi flood control project by justifying it as a relief program for the unemployed.[103] Spending for rivers during Hoover's administration totaled $700 million, with over $150 million for Mississippi River flood control. This enabled the Corps to complete major portions of its navigation and flood control projects on the coasts, the Mississippi and its tributaries, the Hudson, and the Sacramento.[104]

Apart from such spending, Hoover opposed providing direct federal aid to the unemployed. The Republican platform in 1932 called unemployment relief a duty of local government and private agencies. Hoover favored waiting out the Depression. Instead, the economy worsened, and the voters sent Hoover into retirement.[105]

Conclusion

As in the past, lobbyists in 1928 were prepared to politically exploit a major flood. This approach had been instrumental in convincing Congress to create the Mississippi River Commission in 1879, the House Flood Control Committee in 1916, and the Flood Control Act of 1917. The key flood control lobby organization had been disbanded after the 1917 Flood Control Act passed. Even so, the organizational skills and connections built for the 1917 campaign were quickly pressed into use as the Mississippi River flooded. The flood was extraordinary. The legislation that resulted from the new campaign was also extraordinary, as it set out to redesign the country's largest river (the Mississippi) and fastest-rising river (the Sacramento).

Lobbying for relief aid and flood control assistance after the devastating 1927 Mississippi River flood took on the characteristic features of twentieth-century special interest politics. The president's prestige and influence had increased, but decisions still largely centered within congressional committees and involved personal and organizational pressure. The flood control campaign also had the typical elements of campaigns for place-specific development aid. One question that proponents of place-specific aid have to weigh is whether joining with boosters from other areas will improve or impede their chances of legislative success. In fighting for flood control aid in 1928, activists had to consider the value of alliances within their own valleys as well as with other valleys.

While proposals to completely reengineer the Sacramento and lower Mississippi rivers were welcomed by most residents and investors in these two valleys, they soon realized that each engineering proposal would create a different set of winners and losers within each valley. Elites in the Sacramento River largely agreed on engineering options. Proposals to build bypasses had provoked lawsuits by owners in bypass areas, but these were defeated in court. On the lower Mississippi River, however, the 1922 and 1927 floods provoked regional leaders to split. Some levee activists in the Yazoo-Mississippi Delta had opposed downstream outlets as possibly harmful to their upstream area. Elites in New Orleans won this point during the 1927 flood by insisting on their right to force an outlet, but others throughout the lower Mississippi Valley would protest the 1928 plan in later years. Winners and losers within each valley were determined by long-term political and economic struggles, physical conditions, and strategic actions as physical and political circumstances changed.

Advocates in the two valleys also chose to mobilize support from other regions, emphasizing the national economic consequences of the unprecedented 1927 flood. This ran the risk of attracting competition for flood control funds. In 1928, Congress considered and rejected proposals to extend flood control to other regions, but the idea for a national program had been planted. Projects on three other rivers were added the next year. In the coming years, activists in the Mississippi and Sacramento valleys sustained their connections with supporters throughout the country, and they were able to keep their own programs funded when people again made demands for other rivers. In retrospect, the 1927 flood was a unique event but one that was used effectively to argue that flood control as a general function was a national concern. The imprint of federal planners was obvious in the plans that emerged for the Sacramento and Mississippi valleys, but these ambitious plans continued to rely on state governments and levee districts to share some costs and to regulate and maintain some levee lines. Processes from design through implementation affirmed the politics of local and regional deal making and made it clear that army engineering plans had to accommodate these political goals. It would seem natural enough that the 1936 bill establishing a nationwide flood control program would follow the pattern of logrolling long practiced in passing river navigation bills, rather than Progressive Movement ideals of project assessment and ranking.

chapter 10

A Nationwide Program for Flood Control

> It is the sense of Congress that flood control on navigable waters or their tributaries is a proper activity of the federal government.—1936 Flood Control Act

> We are embarking on an $8,000,000,000 program here this morning; that is what I'm driving at.
> —Sen. Josiah W. Bailey, U.S. Senate Committee on Commerce, *Flood Control Act of 1936*, 33

The Special Role of Land Policy in the New Deal

In the 1920s and 1930s, river activists sustained their strong national organization, the National Rivers and Harbors Congress (NRHC), and worked to extend the flood control program to all rivers, a goal that New Deal spending made possible. President Franklin D. Roosevelt and some on his staff pushed to create nonpolitical, professionally managed organizations to plan and supervise multipurpose river development projects. What resulted instead was a piecemeal national system managed by agencies including the Bureau of Reclamation, Tennessee Valley Authority (TVA), Soil Conservation Service, Bonneville Power Administration, and Army Corps of Engineers. With the passage of the 1936 Flood Control Act, the Corps of Engineers retained discretion over most of the country's rivers, and its duties remained focused on engineering rivers for limited purposes, that is, for navigation and for flood control.

The TVA came closest to embodying the grand vision of Progressive and New Deal reformers. It is the only land agency designed to operate autonomously from Congress and the president, and it has the widest purview of any agency over the planning and development of resources in a region. The TVA was created to provide electricity, promote development, and conserve natural resources through enlightened planning. The sponsor of the TVA bill, Sen. George Norris, explicitly contrasted the TVA to the established navigation and irrigation programs of the Army Corps of Engineers and the Bureau of Reclamation. Norris and other New Dealers criticized those two programs for developing resources for single purposes only. They also opposed the tendency of Congress to distribute water projects to many legislative districts, rather than following the cost-benefit analyses and rankings of agency analysts, and they proposed the TVA as an agency that would achieve rational planning.

The TVA has been copied around the world, but Norris and Roosevelt were unable to create additional valley authorities in the United States.[1] This chapter examines why the TVA was created in 1933, why the 1936 Flood Control Act passed, and why subsequent valley authority proposals were rejected. Congress passed the 1936 Flood Control Act with an enthusiastic round of logrolling, empowering the Corps of Engineers to build flood control projects throughout the country and using the projects in the Sacramento and Mississippi valleys as a model for Corps cooperation with state and local authorities. Despite the early promise of the TVA, President Roosevelt's signing of the national Flood Control Act of 1936 effectively ended nationwide, comprehensive planning for natural resources at the federal level.

Assessing New Deal Accomplishments

What accounts for the expansion of the flood control program during a period known for its reforms? The New Deal decisively expanded the state apparatus and increased government intervention into social life. Recent studies by political scientists and sociologists have contended that the New Deal—in combination with earlier Progressive reforms—overcame the limitations of the nineteenth-century central state, created the basic bureaucratic features of the modern welfare state in the United States, helped to shape gender and class distinctions that were emerging in the industrialized economy, and affirmed selected rights for workers and selected powers for labor unions.[2] These researchers find that the federal government intervened into the industrial sectors minimally before the New Deal.

Looking at resource policy, we come to a somewhat different conclusion.[3] As described in earlier chapters, the government promoted land and resource development before the New Deal in the form of court decisions, land grants, and small engineering projects, rather than large, bureaucratic programs. Most New Deal resource policies extended these existing policies, rather than reforming them.[4] Roosevelt agreed to expand numerous programs that originated outside of his circle of New Deal ideologists, even those like the flood control program that New Dealers had targeted for replacement or reform.

The rest of this chapter shows that Roosevelt and the New Dealers in Congress approved both the TVA and the Flood Control Act because they felt the option of taking no action was the least progressive choice. That these two dissimilar river programs were created under a strong president who favored reform points up Roosevelt's pragmatism and the flexibility of the political system. The outcomes of river legislation during the New Deal depended on the status of reformers in electoral politics and on the institutionalization of reformers in government. It also depended on the political status and institutionalization of interest groups.

In river politics during the New Deal, innovators won only when opposing interest groups were weak. Interest groups were critical in shaping the options available to policy makers and in affecting the options they chose. Rather than arguing that either reformist officials or interest group actors were consistently effective in shaping policy, this chapter examines the conditions under which these actors succeeded and the ways that social actors strategized to navigate or shape political structures. Whether they favored valley authorities or the Corps of Engineers, political actors had to coordinate their actions at potential choke points to win. The New Dealers' eagerness to act provided opportunities not only for advocates of reform but also, inadvertently, for opponents of reform.

The New Deal

When the Democrat Franklin Delano Roosevelt (1933–45) took office, no one expected that his administration would intervene so deeply and extensively into public works and other aspects of the economy or that Congress would push for deficit spending and more inflation. As president of the American Construction Council, Roosevelt had pressed state and federal governments to promote economic growth while allowing industries to govern themselves. As governor of New York, Roosevelt had continued to support such policies. But like politicians in other industrialized states, and like Democratic President Wilson, Roosevelt had also provided direct aid to the unemployed. As a presidential candidate, Roosevelt promised seemingly contradictory policies. These included increased spending for unemployment relief and public works and deep cuts in overall spending to balance the budget.[5]

Roosevelt had supported federal spending for flood control, notably in response to the 1927 Mississippi River flood. His special interest was forests, though, not rivers. He usually discussed rivers in the context of integrated resource development plans that featured reforestation of upstream watersheds.[6] Even so, the 1932 Democratic platform was the first major party platform to call for federal flood control projects.[7] As president, Roosevelt confounded categories by combining policies favored by Woodrow Wilson and Theodore Roosevelt with elements of business self-regulation that he had practiced in the 1920s.[8]

The First One Hundred Days

Land development, especially farm relief, was prominent in New Deal legislation from the start. Roosevelt—together with congressional opportunists and allies—abandoned the pledge to balance the budget. He quickly began building what became the New Deal, in a series of pragmatic, ad hoc responses to changing economic, social, and political conditions. During the first one hundred days of the administration, Roosevelt, his advisors, and Congress began to tinker with the currency, set finance regulations, stabilize banks and businesses, and reinflate crop prices.[9]

Officials spent great amounts on infrastructure as well, usually justified for unemployment relief. Roosevelt at first resisted spending on massive public works. Progressive senators and some cabinet members convinced him to add such works to his legislative program. The Public Works Administration (PWA) was created in July 1933 with $3.3 billion to employ laborers. The PWA based its project lists for high dams, waterway improvements, and flood control levees on studies by the Bureau of Reclamation and the Corps. It also directed funding through these and other agencies.[10] But the TVA became the agency that most thoroughly expressed the ideals of rational management of resources so valued by New Deal conservationists.

Opportunity in the Tennessee Valley

The Tennessee River valley was unindustrialized and host to a population considered poor even by the standards of the Depression. Contenders, however, saw these very conditions as political opportunities because the region also featured nearly all of the eastern streams suited for power dams. As TVA engineers would later suggest, the valley's lack of development was a strong advantage. It allowed engineers to build dams without disrupting rail and road systems.[11] These conditions, and unsophisticated lobbying by local elites and interest groups, account for why Sen. George Norris and President Roosevelt were able to target the Tennessee Valley as a site for New Deal experimentation in resource planning.

During World War I, Congress began a program to reduce the country's dependence on nitrate imports for ammunition. Work was begun on a hydropower dam and nitrate plant at Muscle Shoals, Alabama, on the Tennessee River. These projects were not completed in time to produce munitions for the war, but they attracted keen interest from fertilizer producers and power companies. Republicans characterized Muscle Shoals as a boondoggle. Debates soon began over whether the nitrate plant and the uncompleted dam should remain under government control.[12]

The local group most prominent in this debate was the Tennessee River Improvement Association (TRIA). Organized by owners of large farms and by civic boosters to promote local trade by improving river navigation, this group affiliated with the NRHC to lobby for river aid.[13] The executive secretary of the TRIA, J. W. Worthington, had interests in several regional electric companies. He began to promote navigation and power dams for the Muscle Shoals area in 1906 as joint projects of government and business. Worthington apparently convinced army engineers to include power development in their survey of the Tennessee. The TRIA hosted a congressional delegation in 1915. In 1916, Worthington persuaded senators from Tennessee and Alabama to lobby Democratic President Woodrow Wilson to locate the nitrate plant at Muscle Shoals.[14] Despite the valley's obscurity, a series of incidents made Tennessee Valley development into a national issue and exposed the TRIA's heavy-handed lobbying to scrutiny.

Officials in the Wilson administration proposed creating a government corporation to run the facilities or leasing the dam and nitrate plant to industry. Lobbyists played both angles. House Republicans accused the local Muscle Shoals Association of taking secret funding from the Alabama Power Company to print pamphlets urging government funding to complete the dam. The treasurer of the Muscle Shoals Association and president of the TRIA, Claudius H. Huston, denied this. When Henry Ford expressed interest in the dam, Chief Army Engineer Lansing Beach asked Worthington to encourage Ford. Worthington wrote Ford's first proposal. The *Chicago Tribune* reported that Worthington, Chester Gray of the American Farm Bureau Federation, and Claudius Huston of the TRIA, who had become the assistant secretary of commerce, had agreed to seek farmers' support by emphasizing Ford's plan for nitrate fertilizer production.[15]

National Forester Gifford Pinchot and other Progressives opposed Ford. The chair of the Agriculture Committee, George Norris, led this opposition effort in the Senate while his committee considered the Ford bill. Advocates of the Corps of Engineers' river programs, including the Mississippi Valley Association, American Farm Bureau Federation, Mississippi Valley Waterway Association, and TRIA, on the other hand, promoted Ford's proposal. Meanwhile, the *Chattanooga News* reported an unsavory deal by Claudius Huston, the former chair of the TRIA. According to this report, Huston, acting as a new subcommittee official for the Republican National Committee, intended to press Republican President Calvin Coolidge to support Ford's plan and thereby encourage Ford to end his undeclared candidacy for the presidency. When Ford did so in December 1923, and endorsed Coolidge, some accused Ford and Coolidge of striking just such a trade. Politicians, editorialists, and even southern industrialists began to argue that the Ford plan would deliver little to the public. Impatient with political scrutiny, Ford withdrew his proposal in 1924.[16]

Preston Hubbard shows that lobbyists who had been casting about from one proposal to another were neutralized by Norris, who had been inspired to introduce his own bills for the valley from the 1920s on.[17] Norris requested the Federal Trade Commission in 1930 to investigate electric company ties to newspapers that were writing editorials on Muscle Shoals, and Norris publicized the commission's proceedings. A subcommittee of Norris's Senate Judiciary Committee uncovered payments by the American Cyanamid Company to Chester Gray of the American Farm Bureau Federation, for lobbying on Muscle Shoals. Gray had also worked with Worthington and Claudius Huston—who had been chair of the Republican National Committee in 1928—to influence the 1928 party platform planks on Muscle Shoals. And Gray, Huston, and Worthington had lobbied for a Muscle Shoals bill benefiting the chemical industry, for which Huston received $36,000 from Union Carbide.

Republican President Herbert Hoover had seemed ready in December 1929 to support leasing the facilities, but Norris's investigations discredited the bids of electric and chemical companies. Norris passed his bill for government development of fertilizer, power, and flood control projects at Muscle Shoals. He had support from all southern senators, except Louisiana Sen. Joseph E. Ransdell, the past president of the NRHC. Hoover vetoed Norris's bill.[18]

Norris received Franklin Roosevelt's support even before his inauguration as president. Roosevelt encouraged Norris to construct a unified development project guided by social planning.[19] Antimonopolists and social planners in Roosevelt's administration had competing views about how the government should regulate electricity production. Roosevelt unified them in support of his staff's draft bill. Power companies and others joined with conservatives to attack Roosevelt's bill as socialist.[20] Members of Congress were divided over making hydropower production a private or public function.[21] But many voters were dissatisfied with power companies, and most southern Democratic members of Congress favored aid to their hard-hit region. In addition, backers of other development plans remained disorganized and in disrepute. The Tennessee Valley Authority Act passed both houses, and Roosevelt signed it in May 1933.[22]

The Nation's Rivers

The evident willingness of the president and of Congress to approve infrastructure appropriations stimulated existing natural resource agencies and their allies to compete for funds. Advocates of additional valley authorities were at a strategic disadvantage against these organized campaigns. Navigation and flood control advocates promoted those special purpose programs of the Corps of Engineers. The National Reclamation Association lobbied to assign additional rivers and additional tasks to the Bureau of Reclamation.[23] Conservation planners in Roosevelt's advisory National Resources Committee (NRC), who were creating their own designs for natural resource planning, recommended against additional valley authorities.[24] And board members of the new TVA were busy fighting among themselves and uninvolved in lobbying for new valley authorities.[25] No established agency or lobby organization promoted comprehensive planning and development by valley authorities.

The Corps and its supporters had confronted demands for multipurpose development in the past. Most advocates of inland navigation improvement and most army engineers had resisted having the Corps undertake flood control until 1917, let alone multipurpose river development. The Corps had incorporated new activities into its work over time, but usually only after Congress pushed, and only in its own due time.[26] Congress pushed hardest in the Water Power Act of 1920, directing the Corps to survey all navigable rivers for potential navigation, irrigation, and flood control projects. In 1927, Congress began to fund the surveys that the Corps' had published in House Document 308. In preparing the "308" reports, Corps officers broadened their engineering knowledge and contacted a wider range of local officials than ever before. Advocates of other agencies could use the "308" reports to their own advantage, as Sen. Norris did when lobbying for the TVA. But the reports also bolstered the Corps' reputation, which had been damaged after the 1927 Mississippi River flood. Corps officers could now answer critiques that their agency was too narrowly focused.[27]

River Policy Is Delayed

In this institutional and political context, advocates called for expanding the Corps' flood control work to all navigable rivers where needed. The congressional leader calling for nationwide flood control during the 1930s was Rep. Riley J. Wilson (Democrat, Louisiana). He was an early member of the House Flood Control Committee and had served as chair since 1933. Wilson had actively promoted the 1917 and 1928 flood control provisions for the Mississippi River and was president of the NRHC as it lobbied for the 1928 Flood Control Act.[28] Most other flood control advocates in Congress were also Democrats from flood-prone areas. During Roosevelt's first one hundred days, Riley Wilson requested that administration officials include flood control projects in their unemployment relief plans. Wilson received encouragement from the administration, but no concrete promises.[29]

On the other side, Sen. Norris and other senators were the main supporters of valley authorities. Roosevelt spoke publicly in favor of river development and asked Sen. Clarence Dill (Democrat, Washington) to work with Norris to coordinate river valley proposals. He also called a meeting with Dill, Norris, Riley Wilson, chair of the Rivers and Harbors Committee Joseph J. Mansfield (Democrat, Texas), Sen. Hiram Johnson (Democrat, California), and others from the South and West. After the meeting, Norris and Wilson introduced resolutions asking the president to submit a report for comprehensive river development. By the end of 1933, members of Congress had submitted bills for valley authorities in eight major drainage systems. Roosevelt appeared to be working to incorporate the demands for flood control with some form of comprehensive planning.[30]

Contention delayed legislation on rivers, at a time when other resource projects were being approved quickly through agricultural and unemployment relief bills. Roosevelt requested that the cabinet-level Committee on Water Flow answer the resolution from Norris and Wilson. The committee's report called for agency coordination of comprehensive development.[31] Secretary of War George Dern attached a supplement criticizing the committee's arbitrary selection of river basins. He argued that the Corps had already established a priority list and that the Corps could coordinate all but the Bureau of Reclamation's irrigation projects. Roosevelt stood apart from these conflicts.[32]

Meanwhile, Roosevelt organized his own staff to pursue comprehensive resource planning. In June 1934, Roosevelt created the National Resources Board, which became the NRC.[33] Within the NRC, the Water Resources Committee (WRC) began to draw up plans for the Mississippi Valley, with special emphasis on power generation.[34] Roosevelt never clarified the WRC's role in water policy. Absent the mandates and appropriations of operating agencies, these planning boards could only make recommendations to the president.[35] Most legislators were actively hostile to oversight by a presidential board, and so none championed the WRC. Chief Engineer Major General Edward Markham also objected to board oversight and in 1935 began to send aides in his place to attend WRC meetings.[36] The president could have pushed comprehensive planning more vigorously, but he was busy demanding that Congress pass legislation on personal and corporate taxes, labor relations, and social security.[37]

Members of the NRHC commended the president for proposing comprehensive water planning as a program for work relief. They also supported making the NRC a permanent planning commission.[38] This was the first time the NRHC passed a resolution supporting a comprehensive planning agency. Legislators in the NRHC typically wanted the Corps to propose projects directly to Congress, and not through the administration's Bureau of the Budget. They were consistent opponents of the planning boards and their public works proposals.[39]

The Second New Deal

The second period of New Deal legislative activity, in 1935, deepened the federal government's involvement in land development. The president signed a $4.8 million appropriation in April for jobs to provide services or build public works. This was the single largest appropriation by the U.S. government to that time. The Works Progress Administration (WPA), under administrator Harry Hopkins, would undertake small projects and coordinate projects done by other agencies.[40]

This pot of money compelled the PWA, WPA, Bureau of Public Roads,

Department of the Interior and its Bureau of Reclamation, and the Corps of Engineers into a terrific competition.[41] The WPA administered and distributed $11 billion by 1941 to existing and new agencies and employed eight million workers, 20 percent of the workforce in the United States.[42] Construction and conservation projects, including those for swamp drainage and levee protection, constituted 78 percent of the WPA expenditures.[43] Corps officials tried but failed to gain administrative control over water programs that Congress approved under the programs of the PWA and WPA.[44]

Riley Wilson introduced a bill to the House (HR 8455) on 12 June 1935 specifying 285 flood control projects in thirty-four states for $370 million. It was constructed as a traditional authorization bill. If Interior Secretary Harold Ickes and the president refused to fund the projects from the WPA's emergency relief fund, Congress could make a direct appropriation. This bill, with its list of specific projects, was modeled on the rivers and harbors bills for navigation improvements. It became the proposal championed by advocates of single-purpose flood control work by the Corps of Engineers.[45]

Viable Alternatives for Legislation

What alternative proposals for river development were available? What information was available to those on Capitol Hill and in the White House concerning the performance of the Corps of Engineers in flood control management and the costs the Corps incurred in those projects? Legislators and the president appeared ready to approve some form of federal management for the nation's rivers. There were several models at hand, each with political, fiscal, and engineering drawbacks. Environmental drawbacks concerned few at the time.

As time passed, possibilities narrowed, and one outcome for national river policy became more and more likely. Barrington Moore calls these other possibilities suppressed historical alternatives.[46] This section discusses the social and political support that existed in 1935 for the flood control bill and for alternative plans.

Many officials remained enthusiastic about multipurpose river development. The TVA had begun work by 1935 and was attracting local political support as well as international attention for its economic development work.[47] Sen. Norris was the key promoter of valley authorities. At the beginning of 1934, a letter signed by Roosevelt had suggested to Norris that the president would be directing the PWA's Mississippi Valley Committee to undertake comprehensive planning. Sen. Dill wrote to the president that members of Congress were proposing too many valley authorities for their own localities. Congress was likely to drop them all, he noted, unless advocates could compose them into a single bill. In his state of the union speech in January 1934, Roosevelt mentioned that new projects like the TVA were being considered, although he did not commit his support to this policy. The president's advisory WRC was working on river development plans that had not been released.[48] So, the advocates for multipurpose development were mainly legislators and officials of advisory commissions who united behind no single plan and left policy leadership to the president.

Policy makers understood that the political and economic costs of expanding river programs would be high no matter which agency undertook the work, but they had more information about the engineering, political, and fiscal problems of the long-standing Corps programs than they had about other options. By 1935, many of the engineering problems that would eventually arise in building the Corps' Mississippi and Sacramento river flood control projects were evident. The Corps regularly reported these problems in annual reports and in testimony before congressional river committees. Debris control works on the Yuba River—the river most heavily damaged by mining—were completed that year and were to be maintained by both federal and state agencies.[49] The Corps' close consultation with local elites was also well understood.

One political problem for the flood control program was the difficulty that states on the lower Mississippi River were having in convincing landowners to grant or sell rights-of-way in areas designated in the Boeuf Floodway, which was designed to receive overflow only rarely.[50] The Boeuf Floodway would have nearly duplicated the Cypress Creek outlet near Arkansas City, which local interests had closed in 1921 with the approval of the Mississippi River Commission. Landowners in the proposed Boeuf Floodway area argued that they should receive the same compensation for "flowage rights" that owners in other floodway areas were to receive. When President Coolidge had approved appropriations for the other floodways, it became clear that he had no intent to send payments to Boeuf area owners. Owners of many parcels in the proposed Boeuf Floodway responded to the floodway plan with lawsuits and protests, and local members of Congress pressed President Hoover in 1929 to approve the compensation. His attorney general stated that this would require new legislation. Litigation carried on, and the Supreme Court finally decided in 1931 that the government had to compensate owners. Army engineers and politicians in the meantime considered cutoffs and reservoirs as alternatives.[51] In February 1935, the Chief of Engineers submitted a report to the House Committee on Flood Control proposing to build a Eudora Floodway in Louisiana along the Tensas River instead of building the Boeuf Floodway.[52] Both of the proposed floodways were soon defeated by local protests.[53]

One emerging fiscal problem for the flood control option was the trend to assign a greater share of project costs to the federal government. Recent bills had required local and state governments to share levee-building costs, to purchase rights-of-way, and to maintain flood control projects. NRHC resolutions in 1935 had urged the federal government to purchase rights-of-way for navigation and flood control projects on the Mississippi River.[54] More generally, it was apparent that the nationwide flood control program proposed in HR 8455 could eventually cost the federal government billions of dollars.

Roosevelt remained characteristically noncommittal when faced with these competing interests.[55] Roosevelt directed one of his staff members to ask House Speaker John O'Connor and Rep. Riley Wilson why they had proposed the single-purpose flood control bill, which was not in Roosevelt's budget. A delegation of forty-four members of Congress met with the president in July 1935 to encourage him to support Wilson's flood control bill.[56]

Was the situation "open"? Roosevelt treated the situation as somewhat open, although one could also surmise that he delayed announcing his position because he felt it politically advantageous to do so. Joseph Arnold notes that while Roosevelt consistently supported proposals for comprehensive planning and development, he never stated that he would refuse to support less-than-comprehensive proposals in the meantime.[57]

Floods Again

Conservative northeastern politicians had generally viewed the flood control program as a western attempt to loot eastern tax revenues. Major floods may have made it politically feasible for them to fund the program. As in the past, flood control lobbyists, army engineers, and supporters in Congress took political advantage of the situation by pressing their demands. Spring flooding in 1935 in New York, Washington state, West Virginia, and in states in the Midwest and South caused $130 million in damages and killed 236. It also yielded one hundred flood control bills by August. Roosevelt passed along flood relief funds via the Reconstruction Finance Corporation.[58]

In August 1935, the ranking Republican on the Flood Control Committee called Wilson's bill the biggest pork barrel he had seen. Some defended pork, but others, including Republican Dewey Short, the Missouri State vice president of the NRHC, said the bill was vital aid, not pork. The bill narrowly passed the House. By the time senators in the Commerce Committee passed it to the Senate, the bill had half a billion dollars in appropriations, twice the amount in the Senate's entire tax bill. An anti–New Deal Democrat, Sen. Millard Tydings (Maryland), lodged his protest by reading the long list of projects, including "$1,825,000; a mere bagatelle" for the Eudora floodway in Louisiana. The Senate sent the bill back to committee.[59]

While the Senate Commerce Committee met in March 1936, the Potomac flooded the riverfront parks of Washington, D.C. The conservative Democratic chair of the Commerce Committee, Royal Copeland of New York, was impressed by the widespread support for Wilson's bill among editors, members of Congress, and upstate New York activists who had formed a new council allied with the NRHC. Critics were largely silent. The press was now covering Sen. Copeland's public statements about the bill regularly. When members of the Commerce Committee were unable to settle on a list of projects, they asked Chief Engineer Markham to revise the original version of HR 8455.[60]

At its largest convention to date in April, the NRHC resolved in favor of replacing the Boeuf Floodway with the Eudora Floodway and in favor of the flood control bill, HR 8455.[61] NRHC delegates approved of soil conservation and reforestation projects as well, but they emphasized flood control and sought to ensure the Corps authority.[62] The NRHC's 1935 resolutions had recommended that sponsors of flood control projects get them approved by the NRHC Board of Directors. This stance was the reverse of the NRHC's principled pledge to lobby for "a policy, not a project." It also indicated that the NRHC members expected to have the same privileged position as advisors to the secretary of war and to Congress on questions of flood control that they had enjoyed on questions of navigation.[63] Members of the NRHC, the Ohio Valley Conservation and Flood Control Congress, and activists from New England and upstate New York appear to have been the key lobbyists for the bill.[64]

Roosevelt delayed committing to a reelection campaign based on expanding New Deal programs like the TVA. Roosevelt faced certain challenge from populists in the Union Party and from Republicans. He faced a possible challenge from small-government Democrats, who might decide to bolt from the party.[65] As Kenneth Finegold and Theda Skocpol note, Roosevelt's perception of his party base affected his decisions about policy.[66] But working through parties was not the only way that social groups affected policy choices.[67] River lobbyists had criticized the TVA and promoted the flood control program as the best option. By the time that Roosevelt ruled out supporting more valley authorities, flood control was the only option left.

At the end of April, Roosevelt reported to the Senate majority leader, Joseph Robinson, that he found the flood control bill unsound. Leaders of the Soil Conservation Service, the U.S. Department of Agriculture (USDA), the NRC, and the WRC asked President Roosevelt to broaden the bill to include resource planning and erosion prevention. Sen. Carl Hayden (Arizona) met with the president to present an amendment. Evidently written in haste with the assistance of USDA officials, it added soil conservation projects and small upstream reservoirs. Morris L. Cooke of the WRC wrote to Roosevelt in favor of this test of upstream engineering, noting that Sen. George Norris also approved the amendment.[68]

The Senate and then the House passed the amended bill early in June 1936 and sent it to Roosevelt. The bill was a typical product of compromise in Congress, being "fairly clear in its general goals, but confusing and even irrational in its specific policies and administrative machinery."[69] The Republicans and Democrats held their nominating conventions that month. The Democratic platform called for the federal government to coordinate responses to flooding and other problems that states are unable to meet. The Republican platform, written from a position of weakness, urged that public works projects at least be decided on their own merit and kept distinct from relief spending.[70] On 22 June Roosevelt signed the flood control bill.

The 1936 Flood Control Act affirmed that flood control projects "are in the interest of the general welfare," and it established the federal government's responsibility for flood control on all navigable streams. The act designated $310 million for dozens of flood control projects. The Corps would build large downstream projects. The Soil Conservation Service and other USDA agencies would build small headwater reservoirs. Local interests were to pay for land, easements, and damage caused by construction.[71] The term "navigable streams" in practice posed a negligible restriction on the Corps' potential purview. The chief of engineers by that time defined it as any stream where one could float an item of commercial value for any distance.[72]

Political Action, Contingency, and the End of the Alternative Policies

After passage of the Flood Control Act, New Dealers were less active in promoting proposals for valley authorities or other forms of planning. Leaders of the existing natural resource agencies and their supporters continued to oppose proposals that would change their authority. There were two planning models available in 1936. The valley authority model was supported mainly by Sen. Norris and a few other legislators. A variety of planning options were floated by the advisory NRC but never introduced in legislation.[73]

Comprehensive planning options were soon closed off. Proponents of the NRC and proponents of valley authorities were few in number, unable to attract a social base of supporters, and established no firm links with an existing agency.[74] Richard Colignon remarks that because antimonopolists and planners in the New Deal coalition split after 1933, they were unable to champion additional valley authorities against conservative networks.[75]

The alliance of antimonopolists and planners did not have to be especially strong to pass the TVA. This is because flood control activists and other opponents of valley authorities had been weakened at the time the TVA passed. In addition, Roosevelt and Norris acted with unusual skill and determination to promote the TVA proposal very early in Roosevelt's administration, even before Congress had organized a pro–New Deal bloc. Then the flood control activists roared back. Even if Roosevelt had sustained the alliance of antimonopolists and planners, it is difficult to imagine that strong valley authorities could have been created after the 1936 Flood Control Act solidified the Corps' authority. Roosevelt did test the valley authority option two more times, in 1937 and 1944.

Efforts to Revive Comprehensive Planning

In March and April 1937, Sen. Norris and President Roosevelt told reporters that they were working on plans for seven or eight regional valley authorities to cover the entire country. Agriculture Secretary Henry Wallace, who administered the Soil Conservation Service, and War Secretary Harry Woodring, who administered the Corps of Engineers, objected to the plans. Wallace and Woodring told Roosevelt and Interior Secretary Harold Ickes that regional authorities would be fine if they were limited to the planning role. These two cabinet members opposed creating new agencies that would directly administer projects, for fear of demoralizing existing agencies. Planners in Roosevelt's advisory NRC opposed features of the Norris bill that conflicted with their own proposals. Power companies and Sen. Copeland opposed power production by government authorities.[76] Roosevelt had a staff member redraft Norris's bill and draft a competing bill with the watered-down planning features that Wallace and Woodring had suggested. Sen. Joseph J. Mansfield, chair of the House Rivers and Harbors Committee, introduced the second bill. Roosevelt told Secretary Wallace that he would let legislators determine the outcome in committee.[77]

Roosevelt spoke publicly in favor of valley authorities on several occasions, but his speech to Congress on 3 June was "restrained to the point of ambiguity" on the subject of regional authorities. It even appeared to favor Mansfield's approach.[78] Members of Congress who favored the

Corps of Engineers or who opposed government intervention kept the two bills tied up in committees. By November, Roosevelt told Mansfield and NRHC President William J. Driver that he did not wish to create "little TVAS" and would work to pass navigation and flood control bills instead. Congress defeated both the strong and weak valley authority bills.[79]

Planning advocates in the administration continued to discuss resource planning. During World War II, Roosevelt reframed the purpose of valley authorities. He proposed to Congress in 1944 that valley authorities be created for the Missouri, Arkansas, and Columbia valleys as economic development programs to ease the transition once the war ended. To forestall this proposal, Secretary Ickes had Interior Department managers draw up plans to reorganize their own operations into regional divisions, especially focusing on the Bureau of Reclamation. Senators, academics, and journalists promoted the new valley authority proposals. Citizen groups formed in the Columbia and Missouri valleys to back plans for their valleys. Staff members in the Interior Department's Bonneville Power Administration proposed to convert their own agency into a Columbia Valley Authority.[80]

After Roosevelt died in office in 1945, President Harry Truman voiced support for a valley authority on the Columbia, but Truman also supported the Pick-Sloan compromise (discussed below), which divided authority over the Missouri River between the Corps of Engineers and the rival Bureau of Reclamation. The compromise deflated interest in a valley authority for that river. Truman again promoted valley authorities in 1949 after he was elected in his own right. Proponents of a Missouri Valley Authority continued to introduce legislation into the 1950s. Opponents during this period included the Corps of Engineers, the Bureau of Reclamation, the Department of Agriculture, their backers in Congress, activists backing the Bureau of Reclamation, and the NRHC, which organized an opposition coalition called the National and Regional Land and Water Organization's Coordinating Committee.[81]

As the years passed, Truman began to doubt that valley authorities would be implemented in the way that he favored. Opponents kept valley authority bills stalled in congressional committees.[82] Valley authorities never again became a serious policy option.[83] Congress established the Water Resources Council in 1965, but it had little power to coordinate projects.[84] Agriculture, forestry, wildlife, and water policies remained divided among agencies that sustained particularistic relationships with client groups.

Consequences of the 1936 Flood Control Act

Through the years, some legislators opposed creation of resource agencies that would be autonomous from Congress. Other critics attacked them as socialist or even bolshevist schemes.[85] Sen. Copeland fought valley authority bills, because they would undercut the private power industry.[86] President Roosevelt's support for a more powerful NRC had been inconsistent. Few in Congress were interested in creating a permanent agency like the NRC that would intervene between congressional committees and executive agencies. And there were few local supporters of the various proposals floated to create valley authorities.[87] The failure of valley authorities and other planning ventures was a failure of Progressive and conservationist intellectuals and their policy groups.

It is commonplace to note that the New Deal was significant less for the amount of money spent than for the expansion of government duties. This observation holds for New Deal land programs. By 1936, Congress and the administration had restored average spending on public works to only 60 percent of pre-Depression government spending, according to one estimate.[88] New Deal land policies did shift much of the financial and legal responsibility from landowners, and from local and state governments, to the federal government.

Legislation after 1936 continued these trends. In response to a dispute between states over the costs of providing land for reservoirs in the Connecticut Valley, Congress in the 1938 Flood Control Act dropped any requirement that state or local authorities pay for land acquisition, rights-of-way, construction, or maintenance for flood control reservoirs and channel alterations.[89]

Engineering Compromises due to Local Resistance

Pressure from Mississippi River landowners prevented the Corps from completing its revised flood control plan. Landowner protests in

Arkansas had gotten the federal Mississippi River Commission (MRC) to effectively cancel the Boeuf Floodway in 1936. Later landowner protests in Louisiana got the MRC to cancel its replacement, the Eudora Floodway, in 1941.[90] The MRC then abandoned the floodway idea as too costly, given the court costs and the compensation that owners were demanding.

What engineering option remained? The Corps of Engineers had long accepted channel-deepening and straightening as methods for improving navigation, although they considered cutoffs unpredictable. Past Corps publications had denied that channel-straightening would aid flood control.[91] Regardless, the MRC decided to avoid conflict with landowners by deepening the channel and straightening the lower river with cutoffs. They now argued that these steps would increase the river's capacity for large floods.[92] In the 1940s, Congress also began to require that the federal government pay for land where Corps plans required wide setbacks between the river channel and new levees.[93] The government's plan was now designed to protect very vulnerable low-lying lands in the areas once proposed for the Boeuf and Eudora floodways and to direct excess floodwaters through the Atchafalaya region, where political resistance was weaker.[94] Improved scientific methods and the adoption of a wider array of techniques produced a much more effective flood control system on the lower Mississippi.[95] Through its continued vulnerability to local resistance, budget cuts, and engineering debates, however, Corps plans have been repeatedly modified, and many such modifications have increased the overall vulnerability to flooding for the entire lower river area by increasing protections for specific sites.

The Corps vs. the Bureau of Reclamation

With the failure of comprehensive planning, it became clear that the Corps and the Bureau of Reclamation would battle over water resources. The 1938 Flood Control Act allowed the Corps to operate power facilities at flood control reservoir sites.[96] The Reclamation Project Act of 1939 made it official policy for the Bureau of Reclamation to provide electricity as well as irrigation.[97] California's Central Valley and the Missouri Valley became the sites of the most heated competition between the Corps and the bureau.

Large landholders at the southern end of the Great Central Valley, far south of the Sacramento Valley, had congressional representatives add to the 1944 Flood Control Act a Corps dam project on the Kings River. Under reclamation laws at that time, farmers could benefit from the Bureau of Reclamation irrigation programs only if they held 160 acres or less.[98] On the other hand, the 1944 Flood Control Act was being written to allow the Corps to provide water for local irrigation use at no cost, provided that the Interior Secretary approved.[99] This group of farmers succeeded in circumventing the acreage limit by getting a dam built by the Corps instead of the bureau. Thus began a rivalry between these two agencies for control over the Kings River and other California rivers.[100]

The pattern of close cooperation between local land and development interests and federal land agencies produced an even greater conflict in the Missouri River Valley. Some cities on the river struck alliances with the Bureau of Reclamation, some with the Corps of Engineers. After considerable conflict, Congress cobbled together the two plans into the makeshift Pick-Sloan Compromise in 1944. This legislation initially approved 316 separate projects. The Corps would manage dams, navigation, and flood control in downstream reaches. The Bureau of Reclamation would manage upstream reservoirs for irrigation projects, municipal uses, and power production.[101]

Policy Networks in River Policy

The Corps, congressional river committees, and the NRHC continued to work as a policy network for river management. The NRHC continued to attract its usual constituency of large landowners, local and state officials, and representatives of trade associations, officers of the Corps of Engineers, and members of Congress. Critics say that its work became more secretive between the 1930s to the 1960s. The NRHC would review plans for navigation and flood control improvements proposed by the Corps, often even before the Corps sent its recommendations to congressional river committees.[102]

Congress dramatically expanded the Corps of Engineers' work on inland waterways in the 1940s, adding dozens of rivers to its roster. The massive 1944 Flood Control Act authorized the Corps to add park and recreation areas at reservoirs and allowed the Corps to subsidize its work by selling water to municipal and industrial users.[103] In response to persistent agitation by sport fishers and hunters, Congress in 1946 provided for the protection of fish and wildlife at flood control project sites.[104] By the 1950s, congressional generosity in navigation and flood control improvements provoked yet another generation of critics of the pork barrel.[105]

Challenges to the River Issue Network

Declining employment in agriculture and increased evidence of river and lake degradation changed the tight connections between congressional subcommittees, the river lobby, and the Corps of Engineers. Hunters, fishers, wildlife enthusiasts, and environmentalists began to pose serious challenges to the engineering policies of the Corps and other resource agencies. The 1944 Flood Control Act expanded the definition of flood control to include major land drainage projects, including "nonstructural alterations" in the use of floodplains.[106] In the 1950s and 1960s, Congress required the Corps to adopt floodplain management techniques developed by the TVA. The Corps was to shift its emphasis from flood prevention, which it had never achieved, to the reduction of flood damage.[107] The 1969 National Environmental Policy Act required the Corps to undertake a broad range of environmental impact studies.[108]

The chief objects of environmentalists' scorn in water policy through the 1970s and 1980s were the high dams for water and power supply built by the Bureau of Reclamation and the Corps of Engineers.[109] These dams flooded river canyons, sterilized portions of the inland waters, disrupted fish mating runs, and diverted water from many natural areas. The Corps also attracted criticism for directing stormwater so quickly to the sea, through concrete-lined channels and culverts. This design reduces a river's average volume, so that salts become concentrated in the river. In areas with intensive flood control systems, local water agencies often have to divert other sources of water for agricultural and municipal use, at great environmental and economic cost.[110]

As competition over water resources has increased, new claimants have gained some influence in decisions. The Water Supply Act of 1958 directed the Corps to provide storage in its reservoirs for municipal and industrial water supply, if local users requested this and paid the costs.[111] This order stands in direct conflict with the need to keep reservoirs empty to receive flood waters. Members of congressional committees—and officials of the Corps and the Mississippi River Commission—now consult organizations representing environmental activists, commercial fishers, crawfishers, shrimpers, oyster gatherers, recreational hunters and others, besides consulting city water authorities, levee districts, and shipping companies.[112] And once again, the Corps has managed to accept a new duty, that of reversing some of the engineering works that it built years ago, so as to restore wetlands.

Not surprisingly, changes to water institutions have been slow, although budget cuts since the 1970s have forced some changes. Water projects have always been conspicuous forms of discretionary spending, in the late twentieth century, constituting about three percent of discretionary federal spending. While President Ronald Reagan's professed interest in cutting budgets often resulted in a redirection of spending to contractors, it did yield cuts in federal water projects that appear to have become permanent. In the past, advocates of irrigation, hydropower, navigation, and flood control might have exchanged political favors to win more overall spending on infrastructure, but since the 1970s they have generally fought against each other.[113] Large water lobby organizations have also lost their once unassailable dominance as shadow government organizations.[114] Even so, a host of pro-growth pressure groups, ranging from local chambers of commerce to shipping organizations, still lobby their local legislators. And at base, the funding mechanism remains the same. Members of Congress still have an appetite to host new projects in their home districts.[115] Present-day complaints from journalists and public interest groups about the Corps of Engineers' secrecy, and government investigations into its accounting practices, reprise criticisms voiced in the 1950s.[116] Environmental restoration projects led by the Bureau of Reclamation and the Corps of Engineers may turn out to be subject to the same problems as the engineering projects of the past.[117]

Persistent Problems in River Management

Water engineers and policy analysts have identified features of the United States' political structure that block effective land use planning in flood control projects. Lack of coordination has led to increased damage costs, even as more projects are built. Floodplain use is under the purview of local and state governments, which historically have imposed few restrictions on property owners. Congress has approved federal flood control systems without requiring that local and state governments restrict building on nearby floodplains. These lands remain vulnerable to flooding even after levees are built.[118] What is more, many residents and farmers who live or work in floodplains—and even in emergency floodways—in turn demand additional federal levee and pumping projects.[119]

The 1986 Water Resources Development Act specified that local and state governments would share the costs of most new Corps projects. Much of the work involving federal, state, and local agencies, however, continues to center on distributing federal funds.[120] Martin Reuss argues that requiring local and state governments to share in managing and paying for water development marks the 1986 act as a step toward restoring relations among federal, state, and local governments that once existed in the nineteenth century.[121] The pattern articulating management among the three levels of government persisted even during the post–World War II period of big federal spending. It reemerged in new form, with regional level governmental initiatives and with nongovernmental organizations and contractors taking on stronger roles, as environmental concerns and problems of inequity among social groups have been addressed by water management laws and institutions. It is unclear whether responses to hurricane damage on the Gulf Coast in 2005 will shift most of the cost of river and hurricane levees back to the federal government for New Orleans, for all damaged Gulf Coast areas, or even for major flood control projects across the country.

By the end of the twentieth century, the federal government had become obliged to respond to large floods and to other great and sudden emergencies. And since 1973, the federal government has, in effect, subsidized construction in floodplains through its National Flood Insurance Programs. The program's payouts to victims are too low to cover much of their losses, but its existence may lead landowners to assume that life on the floodplains involves risks that can be estimated and recovered.[122]

Despite new environmental requirements, the Corps' devotion to building structures has been difficult to reform.[123] One Corps report notes that the agency provides assistance for floodplain development to local residents and governments "on request."[124] Voices for environmental reform within the Corps remain faint.[125] Environmental restoration projects are usually undertaken as smaller, supplementary works. Many are completed simply to fulfill the specific requirements of environmental regulations, rather than being integrated as goals for waterways stewardship. The Corps does undertake more massive restoration projects, notably its project to restore water flow into the Everglades, but these larger programs are, at base, another set of engineering works.[126]

The Corps' inland waterway programs have become the most costly and extensive infrastructure systems in the United States. Taken by itself, the flood control system is surpassed only by the federal highway system in cost.[127] The Corps has permanently changed every major river in the United States and is responsible for draining more of the country's wetlands than any other single organization. And yet most of the time, these programs are invisible to the public, financially and even physically, because the projects have made rivers seem predictable.

Fully Designed Rivers

The Sacramento and San Joaquin valleys have been utterly transformed. Public and private projects for navigation improvement, flood control, swamp drainage, and irrigation have changed the Great Central Valley from a land of alkali flats (in the southern region) and swamps (throughout the valley) into the most intensively farmed valley in the United States. Elna Bakker considers the valley to be the area most radically changed by white settlement in the United States.[128] Mountain mining areas remain heavily scarred and polluted with mercury and other chemicals used to extract the gold. Mining debris remaining in the mountains has formed permanent terraces along mountain streams.[129] The 1928 Sacramento Flood Control Project was judged 90 percent complete by 1944 and complete in 1968, including levees and the Yolo and Sutter bypasses. State and local governments have assumed routine maintenance of these works.[130] Under separate state and federal projects, the Sacramento Deep Water Ship Channel, mountain dams built for hydropower and water supply, and pumps diverting delta water for farm irrigation have also changed the river system, in ways that often conflict with flood control goals. What is more, of the six thousand miles of levees in these valleys, only one-third of these levees meet current federal standards, and the other two-thirds have suffered repeated breaks.[131] Tracts of drained and reclaimed agricultural land in the Sacramento-San Joaquin Delta have been abandoned in recent years as the costs of maintenance rise.

Historian Robert Kelley remarks that the Sacramento River project "was conceived and designed to protect farmers, and now it is having to protect large urbanized metropolitan areas."[132] As a consequence, a levee break can be terrifically costly. The region's water cycle, from winter snow conditions to spring floods, is also proving to be much less predictable than engineers had thought it was. Floods in 1955, 1986, and 1998 exceeded engineers' predictions, often with disastrous results.[133] Competition for the water has been a critical issue in conflicts over California's future growth. Persistent problems include the intrusion of environmentally harmful saltwater into the Sacramento-San Joaquin Delta due to water diversions, declines in fish populations, conflicts between water users in urban and rural areas, and conflicts between water users in northern and southern California.[134]

The Mississippi River is now the "government river."[135] North of St. Louis, it is a series of ponds between navigation locks. South of Cairo, it is freed of obstacles, deep, and easy to navigate. The Corps of Engineers has built a straighter channel by cutting off river bends, especially near Greenville, Mississippi. The banks of the main channel are stabilized and topped with levees. This causes the river's silt to pass by the lands it used to fertilize and to settle in the Gulf of Mexico. The area of the Gulf where the river empties its silt, laden with agricultural chemicals, is now an ecologically dead zone. The Louisiana coastline, with its protective wetlands built through millennia of silt deposition from the Mississippi and Atchafalaya rivers, is meanwhile eroding steadily and increasingly vulnerable to hurricanes. Even with all of these changes, and after spending over $9 billion in federal funds, the 1928 flood control plan remains unfinished.[136] Of the 1,600 miles of levees on the main river stem, 304 miles remain below the federal standard. This problem is more crucial on the Mississippi than on other rivers, because the levees concentrate an enormous volume of water.[137]

The main river channel levees south of Cairo have remained intact since the 1927 flood. The Bird's Point–New Madrid Floodway and the

West Atchafalaya and Morganza floodways (redesigned in various ways since originally proposed) offer potential outlets for floodwaters, as does the Bonnet Carré Spillway. Reservoirs and weirs also control the flow of many tributaries into the Mississippi. Increased land drainage throughout the Mississippi Valley, improved levee containment, and closure of river outlets that formerly emptied into swamps, however, have continued to increase the river system's volume. Catastrophic floods have occurred on the river's tributaries, far larger than those predicted by the Corps of Engineers.[138] The 1993 floods on Mississippi River tributaries caused an estimated $12 billion in damage, leaving 70,000 homeless and killing fifty.[139]

The fortified lower river bears the potential for a grave disaster because of the tremendous volume the river carries, but various interventions of the past also have the potential of leading the Mississippi west, rather than east. The largest remaining outlet is the Atchafalaya River, which links the Red and Mississippi rivers. The Corps over the years has built extensive levee systems along the Atchafalaya to compensate for a series of ill-fated cutoffs, diversions, and blockages intended to straighten the Mississippi's navigation channel near its outlet to the Atchafalaya River. These efforts inadvertently and unexpectedly increased the Mississippi's flow into the Atchafalaya.[140]

As early as the 1920s, some Mississippi River Commission engineers suggested the Atchafalaya might eventually draw the entire Mississippi River into its channel, if not controlled. The MRC officially rejected this scenario, but finally accepted it in the 1950s. Nineteenth-century solutions for a local navigation problem have, therefore, yielded the much bigger problem. At stake is whether Baton Rouge and New Orleans will continue to receive a navigable river. One of the mammoth Old River Control Structures controlling the flow into the Atchafalaya was severely undermined by a 1973 flood. It is now all that keeps the Mississippi River from shifting into the Atchafalaya and making New Orleans a salt-water port with inadequate depth for shipping.[141]

The flooding and evacuation of New Orleans and nearby Plaquemines Parish after Hurricane Katrina will likely become a cultural and policy milestone similar to the 1927 Mississippi River floods. Although flooding occurred through breaks in drainage canals and in hurricane levees on a shipping canal, rather than from a failure of Mississippi River levees, media reports from the start explained that these areas had been made vulnerable through their dependence on river levees.[142] Fewer reports noted that the lower portions of Louisiana are sinking because the rivers have been prevented from depositing silt. Experts and politicians began floating ideas for engineering projects, much as they had after the 1927 river flood. The competing plans pit upstream against downstream (in this case New Orleans against coastal wetlands regions), and will provoke conflicts between shippers, environmentalists, fishers, shrimpers, and oil and gas extractors. The federal government response will likely include silt pumping to portions of the wetlands, storm surge barriers at Lake Pontchartrain, and ever bigger levees and floodwalls. No amount of engineering, however, will make this area safe.

Conclusion

Sociologists have treated the New Deal as an extraordinary period that defined the central state's basic powers to intervene in the economy. This chapter shows that although the New Deal increased the funding and scope of land programs, New Deal programs did not manage to directly change the social character of state intervention into land use. Instead, the New Deal reaffirmed relations between the state and landowners that had been developing since the nineteenth century. In Timothy Mitchell's terms, the New Deal did not radically shift the perceived border for land policy between an authoritative state and society.[143] Increases in the scope of federal flood control in 1928 and 1936 certainly shifted discretion over some river features from local and state levels to the federal level, but implementation still relied on the articulation of the Corps of Engineers with state agencies and with landowner-controlled levee districts. Individual property owners would lose battles over project design, but the flood control program continued to affirm intensive land uses and the general rights of property owners. On the other hand, massive New Deal spending opened the way for much more intensive manipulation of rivers. The indirect effects of this manipulation on the border between state and society are still unfolding.

By considering social displacements of the New Deal period as the consequences of changes in capitalism and in state rule, we can interpret the actions of river activists. Organized business and farming concerns in river valleys participated in decisions that exempted farm labor from federal regulations. They also pressed the central government to reduce some of the environmental uncertainty of their operations through resource programs like the flood control program. Franklin Roosevelt adopted projects that had been shaped by policy making structures outside of his control, structures built before the New Deal.

chapter 11

Rivers by Design

Managing Rivers, Past and Present

Over the years, rivers have had many political meanings and many economic uses. Only some of them are relevant to us now. The Mississippi and Sacramento rivers were vital highways to the interior, sources of great fortunes, and symbols of regional identity. They were transformed politically into national rivers. Now they are simply two links in a national transportation system that has declined as road and rail systems have grown. Recreational boaters rather than barges now dominate portions of these two rivers. Farming along these rivers is lucrative for farm owners in some areas but not in others, and as always, it provides a poor living for farm laborers. Gold mining now centers in the open-pit and underground mines of Nevada and South Africa, not in California. Suburban development is the fastest growing form of land use in the Sacramento Valley. Although the Mississippi and Sacramento rivers remain important to regional identity, they no longer dominate their regions' economies or politics.

Despite such changes, we are culturally, physically, and politically constrained by past decisions about these and other rivers. Government river projects have made some uses possible and other uses difficult or impossible. Concrete-lined flood channels prevent the natural resupply of groundwater. Levees deprive wetlands of water and prevent rivers from depositing fertile silt alongside their channels. Political expectations and political constituencies have been built around such river projects and in turn sustain those projects and resist new social and environmental goals.

The United States built an unusually intensive program for flood control to protect its small proportion of floodplains. In most countries, ambitious government officials typically initiate large infrastructure programs such as this. In the United States, regional elites were the ones who called for a national flood control program.

Making the local and regional problem of flooding into a national concern produced several related changes in the institutions that manage rivers. Flood control had always been a mix of private and public duties. Under the federal government's flood control program, private responsibilities for levee building and maintenance were reduced or shifted entirely to the government. Much of the authority over levee standards and responsibility for costs has also been shifted within the government, from local and state levels to the federal level. Responsibility for the overall design of rivers—a function that was unimaginable in the nineteenth century—was created and taken up by the federal government.

The flood control program extended the central government's reach into social life, but it remains responsive to local demands through an implementation system that articulates the work of the Corps of Engineers with state and local agencies and with landowner beneficiaries. In surveying research on water policies, historian Robert Kelley observes that Congress usually wrests ultimate control from experts in the executive branch and that state and local governments retain considerable authority over water policy, even when the federal government steps in. Because these institutional factors exist in tension in our political culture and under the federal system, the nature of specific water policies therefore vary.[1] Kelley might well have also mentioned the tension between the state and private property, an undercurrent in his own work as well as the work of the environmental historians he surveys. Considering these cultural and institutional tensions, this concluding chapter summarizes findings about the social origins of the linked institutions that created and manage the federal flood control program.

Flood Control and the Two Conflicts over State Building

Using Stein Rokkan's conception of state building, the preceding chapters analyzed how the flood control program was built. In Rokkan's view, the rise of the modern state usually provokes two types of conflicts. One is between the centralizing state and peripheral regions. The second is between industrialists and agrarian landowners.[2] The processes leading to state centralization are not necessarily directed from the center. Landowners, shippers, merchants, and financiers in the Mississippi and Sacramento valleys created flood control campaigns in the effort to improve their regions' chances in the industrializing national economy and to shape the way their regions would be politically and physically integrated under U.S. rule.

Regarding Rokkan's conflict over the centralization of political power, regions recently admitted to the Union had to establish state governments and to negotiate their political relationships to national government bodies, especially Congress. Single-minded and persistent organizing and politicking for flood control aid helped politicians to direct federal resources to their home regions and to gain some sense of control over relations with the central government.

Rokkan's second type of conflict emerged only briefly as a head-to-head political battle between farmers and industrialists, during the period when Sacramento Valley farmers sued large-scale gold mine operators over damages. More importantly, this second type of conflict was manifested in both valleys as a struggle to make local agriculture more competitive in an industrializing economy. Those who depended on the agricultural trade in the trans-Appalachian West competed for transportation improvements against the manufacturing states of the Northeast, and each agricultural region competed for market share against other agricultural regions. In the view of activists, flood control aid would improve regional navigation conditions and would enable farmers to intensify production on fertile lands. Farmers in California managed to use government aid to create a new, industrialized form of agriculture, while farmers in the lower Mississippi River area did not. The history of flood control activism suggests that each of Rokkan's two conflicts unfolds as a struggle over the way the state's claimed space is politically and physically organized and linked.

The Nation and Its Distant Territories

In debating river development, state officials and regional elites asserted various ideas about the nation to justify state efforts to control territory. Although demands for regional development aid might appear to us to be self-serving, nineteenth-century petitioners who argued that such aid was in the national interest expressed concerns shared by many across the country that the West be quickly settled and claimed as part of the United States. As Peter Sahlins shows, the residents of regions distant from the seat of central state power may strategically assert their allegiance to the nation even before state officials have troubled to impose territorial controls such as border markers and patrols against smugglers.[3] In this fashion, flood control activists argued that the federal government should improve the Mississippi and Sacramento rivers to secure the national good and to unify the country. Elites in the Mississippi and Sacramento river valleys viewed the northeast's disproportionate share of river improvement aid as proof that the central state was not fulfilling its proper role in managing the trans-Appalachian West. They represented flood control aid for these two rivers as a national duty because the nation's runoff contributed to the Mississippi's floods and because the nation's wealth had increased as a result of gold mining activities that harmed the Sacramento.

The Border between State Territory and Private Property

As ideas about national duties for resource development changed, the apparent border between state and society also changed. In natural resource policy in the United States, Rokkan's two conflicts involved defining the border between the central state's responsibilities and landowners' responsibilities in managing land. As Timothy Mitchell notes, the state appears to stand apart from society, for instance, as an entity that provides for interstate transportation and that regulates land use and property rights. Because this appearance helps to create the aura of authority, Mitchell proposes that we study how spatial organization and other social practices create the appearance that the state is an authoritative entity separate from society.[4]

In the case of flood control, the central state's authority over resources was repeatedly redefined within the context of successive crises in the central state's rule in the West. At the same time, politicians and activists attempted to make each new definition of federal government authority appear to be a legitimate and natural interpretation of private property rights and of the Constitution's definitions of proper federal government powers. In the interest of stimulating local investment and funding state government levee building without direct federal appropriations, swampland grants placed parcels from the public domain into private hands. In the course of investigating the possibility that floods threatened navigation, which was clearly a federal duty, early federal river surveys identified high-value private plots that should get the greatest protection from floods. To preserve the Mississippi River's navigation channel, army engineers rebuilt levees during flood emergencies. To preserve two national rivers, the Corps of Engineers planned and directed projects to redesign those rivers. To ensure shipping, public safety, and ecological functions of the nation's rivers, the Corps now builds projects for navigation and flood control and regulates wetlands on private lands.

These changes in resource policy involved changes in ideas, law, organizations, and infrastructure, and each of these elements of human activity redefined the border between state and society. The idea of the nation and of the government's national duty expanded. Statutes for river development honored property rights and defined new government responsibilities. Implementation plans articulated Corps of Engineers field teams with local and state government agencies, including levee districts, which are run by local landowners but are organized for a public purpose and are generally chartered by subnational state governments. And the country's rivers represent physical borders between the nation's waters, held in stewardship by the state, and private lands.

Land Policy and the Pork Barrel in a Federal System

The federal system was designed to leave most decisions about land use to the local government level. Under a system that elects legislators by district and state, politicians and local elites have also been motivated to promote federal and subnational state-level development policies that would benefit their localities. Seeking federal aid for flood control was done in the name of local and regional empowerment, but it had the additional effect of deeply involving the state and federal governments in decisions about land use.

Most conventional political analyses of federal aid programs consider their effects on core aspects of the central government, especially on bureaucracies and spending. These analyses are typically less concerned with the effects of aid programs on the government's relationship to society. Looking outward from the center, these analysts often argue that place-specific programs like flood control are primarily wasteful pork-barrel programs. Such programs encourage national politicians to focus on parochial interests and draw resources away from central government activities that these critics argue would benefit the wider public. The Army Corps of Engineers' navigation and flood control programs have especially provoked such critics.[5]

Political scientists Robert Stein and Kenneth Bickers respond to the critics with data from the 1980s showing that distributive programs do not drive federal budget increases, that delivering pork to one's home district does not guarantee a legislator's reelection, and that aid programs are not always made into universal benefits.[6] Stein and Bickers also argue that budget size is a poor measure of the political significance of a program. Rather, pork-barrel programs constitute a vibrant sector of the government. Agency staff, members of Congress, and various interest groups create and sustain such programs as they pursue a set of related organizational and political objectives.

The flood control case highlights the implication of the Stein and Bickers findings for federalism. The most significant political outcome of river programs is the set of working relationships they create among federal, state, and local governments and among government organizations, contractors, and project beneficiaries. These relationships were not produced by the deliberate effort of federal officials to co-opt local government authorities. Nor were they produced by the deliberate effort of local elites to create elaborate links to the federal government. They were a by-product of increased government intervention over time. Close working relationships among the river lobby, the Corps of Engineers, and congressional subcommittees were not an inevitable outcome of river policy, and close relations have not been produced uniformly across federal resource programs.[7]

Local Influence and Environmental Reform

Environmental goals fit poorly into this system for distributing projects. Goals for economic development and the protection of property and lives that were set in the nineteenth and early twentieth centuries are human-centered goals that could be met within the framework of decisions made district by district. Goals for environmental protection have recently been added to the flood control program. These goals would reduce some of the immediate benefits for humans. They would also work best, according to ecologists, only if plans were coordinated across the boundaries of localities or congressional districts, a goal that is often attempted but often frustrated.

It is also politically difficult to justify spending for nature-centered goals, rather than human-centered goals. Environmentalists, local residents, and resource extracting companies have recently convinced Congress to redesign several flood control projects, with aims that include protecting the dwindling Louisiana coastal wetlands and restoring water to the Everglades. They are attempting at once to redesign river hydrology, change federal and state government bureaucracies, counter the protests of current project beneficiaries, and reduce local politicians' incentives to attract construction funds intended to make rivers into aqueducts. They face strong resistance. Critics say that the Corps of Engineers and beneficiaries of existing flood control projects have indeed managed to turn these environmental restoration projects into another set of engineering plans. The flood control program remains at base a land development program, despite the addition of goals for public safety or ecological restoration. The amenity values of green landscapes and trout-quality streams are perceived as being vital to the local economy only in some communities. More typically, those seeking to restore environmental functions must overcome arguments that local economies will suffer if ecosystem needs are placed above human needs or even if new human needs are prioritized.

Environmental problems are also difficult to solve across political boundaries. Because the health of each reach of a river depends on conditions throughout the river system, restoration projects done in isolation often fail. In a federal system, the central government could be mobilized to coordinate action across localities. Progressives and New Dealers had aimed to do this through comprehensive water planning, although they focused on human needs, not ecological needs. Those plans largely failed. Instead, the practice of distributing federal projects by district was perpetuated, a practice that discourages the coordination of environmental problems across political boundaries by providing new incentives to pursue local interests.

It remains to be seen whether local residents and army engineering teams designing and reviewing river projects will create innovative and effective techniques for improving ecosystem functions and meeting human needs. Ensuring local control or influence over decisions has appealed to people across the political spectrum in the United States and has been promoted as an avenue for reform. But local control has produced a mixed legacy. The frequent emergence of local power blocs has inspired critics to argue that devolving power to the local level will not guarantee democratic processes or inspired solutions.[8] Reformers may emerge at the local level in some communities to counterbalance local project beneficiaries, but the practice of pursuing pork for the home district and the long-standing resistance to planning by experts will make it difficult for reformers to change priorities of the flood control program as a whole.

It is clear, however, that politicians, bureaucrats, beneficiaries, and environmentalists expect the U.S. government to continue to manage rivers but to coordinate management with other levels of government and with local interests. The difference is that now, the government is responsible not only for economic development and human safety, but also for ecological functions. Even if these new goals prove elusive, the addition of nature-centered goals to the government's human-centered goals could mark a major shift in expectations about the boundaries between state and society.

Appendix 1

TABLE 2. MISSISSIPPI VALLEY RIVER IMPROVEMENT CONVENTIONS

Year	Convention and Place of Meeting	Purpose and Details	Sources
1842	Rivers and Harbors meeting, Cincinnati	Memorialize Congress to increase transport aid for Mississippi River system.	Barry, *Rising Tide*, 34; Williams, "The Background of the Chicago River and Harbor Convention, 1847," 226
1844	Memphis		Barry, *Rising Tide*, 34
1845	Mississippi River Improvement Conventions, Memphis	First interstate conventions. July meeting draws from six states; November meeting has 500 delegates from twelve states. Chair John C. Calhoun, formerly a strict constructionist, advocates federal improvement for the lower Mississippi River. His southern chauvinism polarizes politicians.	MRIC, *Official Report*, 12; Williams, "The Background of the Chicago River and Harbor Convention, 1847," 226
1846	Memphis	Representatives from every southern and western state and Territory of Iowa.	Childs, *Mighty Mississippi*, 83

1847	River and Harbor Convention, Chicago	"The most numerous delegated convention ever held in this country." Over 10,000 delegates from the Old Northwest and upper Mississippi River regions. In part a reaction against Calhoun's strategy for economic independence from the North, in part a reaction to competition from railroads. Precipitated by Polk's veto of 1846 Rivers and Harbors Bill.	Delegates from the City of St. Louis, *The Commerce and Navigation of the Valley of the Mississippi and Also That Appertaining to the City of St. Louis*; Hall, *Chicago River-and-Harbor Convention July 5th, 6th, and 7th, 1847*; Harbor and River Convention [of 1847] Chicago, *Memorial to the Congress of the United States of the Executive Committee of the Convention Held at Chicago, July 5, 1847*; Harbor and River Convention Chicago, *Proceedings of the Harbor and River Convention Held at Chicago, July Fifth, 1847*; Williams, "The Background of the Chicago River and Harbor Convention, 1847"; Williams, "The Chicago River and Harbor Convention, 1847"
1850	Chicago		Childs, *Mighty Mississippi*, 87
1850	Evansville, Indiana		Childs, *Mighty Mississippi*, 85

1851	Rapids Convention, Burlington, Iowa	Three hundred delegates from the region seek aid for removing obstructions at Des Moines and Rock River rapids.	Childs, *Mighty Mississippi*, 85; Drake, *The Mission of the Mississippi River*; MRIC, *Official Report*, 13; Rapids Convention [of 1851] Burlington, *Proceedings of the Rapids Convention, Held at Burlington, Iowa, on the 23rd and 24th of October, 1851*
1860s	Mississippi Valley Commercial Convention, Keokuk, Iowa	Protesting unfair railroad competition against river transport.	Childs, *Mighty Mississippi*, 121–22
1866	Mississippi River Improvement Convention, Dubuque, Iowa	Seek aid to improve navigation of rapids on upper river. Organized by Merchants' Exchange of St. Louis, 500 attending.	MRIC, *Official Report*, 14; MRIC Dubuque, *Proceedings*
1867	Keokuk, Iowa	Seeking federal improvement of rapids on upper Mississippi River.	MRIC, *Official Report*, 15
1867	River Improvement Convention, St. Louis	Called by Union Merchants' Exchange of St. Louis. Over 300 delegates from states bordering Mississippi River, seeking improvement of rapids and channel improvement on upper river.	Barry, *Rising Tide*, 70; MRIC, *Official Report*, 14; River Improvement Convention St. Louis, *Proceedings*
1869	Louisville, Kentucky		Barry, *Rising Tide*, 70

1869	New Orleans		MRIC, *Official Report*, 15
1872	St. Louis		MRIC, *Official Report*, 15
1873	Congressional Convention, St. Louis	Improve Mississippi River water transport to sea. Governors of Virginia, Missouri, Minnesota, Ohio, Kansas, and 100 members of Congress attend, as well as several thousand delegates.	Barry, *Rising Tide*, 61; Congressional Convention St. Louis, *Proceedings*; MRIC, *Official Report*, 15
1874	Levee convention, New Orleans	Sent a committee to the capitol to lobby Congress for federal support for levee building.	Sitterson, *Sugar Country*, 289
1875	New Orleans		MRIC, *Official Report*, 15
1875	Vicksburg, Mississippi		MRIC, *Official Report*, 15
1876	New Orleans		MRIC, *Official Report*, 15
1877	Mississippi River Improvement Convention, St. Paul	Delegates from boards of trade, chambers of commerce, and other organizations from eighteen states seeking to remove obstructions from St. Paul south. Delegates from upper Middle West.	MRIC, *Memorial*; MRIC, *Official Report*, 15, Sitterson, *Sugar Country*
1879	New Orleans		MRIC, *Official Report*, 15

1879	River Improvement Convention, Quincy, Illinois	East-West sectional inequalities persist; protest that federal projects in West have received only partial funding.	MRIC, *Official Report*, 15; River Improvement Convention Quincy, *Memorial*
1880	New Orleans		MRIC, *Official Report*, 15
1881	Mississippi River Improvement Convention, St. Louis	Delegates from boards of trade, cotton exchanges, and other organizations, politicians. Flood control and navigation improvement of Mississippi River by federal government advocated. Establishes a series of "Western Waterways" conventions.	CIWW New Orleans, *Proceedings*; CIWW Memphis, *Proceedings*; CIWW Washington, *Proceedings*, 76, MRIC, *Official Report*
1881	River and Canal Improvement Convention, Rock Island, Illinois	Prepare for Davenport convention.	River and Canal Improvement Convention Davenport, *Proceedings*, 3
1881	River and Canal Improvement Convention, Davenport, Iowa	Seeking federal aid for canal from the Illinois River at Hennepin (connecting to Chicago) to the Mississippi River; would enable northern water carriers to compete against lower Mississippi River carriers and against railroads. Officers from upper Middle West, New York, and California.	MRIC, *Official Report*, 15; River and Canal Improvement Convention Davenport, *Proceedings, 1881*

1884	Convention for the Improvement of the Mississippi River, Washington, D.C. (in the "Western Water-ways" series)	The Executive Committee on Improvement of the Western Water-ways, formed in St. Louis meeting in 1881, calls this and later conventions. Seeking federal aid for flood control and navigation on the Mississippi. Delegates from 16 states, most from the upper Middle West, but many from the south. Delegates visit President Chester Arthur.	CIWW New Orleans; *Proceedings*, 5, 39; CIWW Memphis, *Proceedings*, 84; CIWW Washington, *Proceedings*
1885	Convention on the Improvement of Western Water-ways, New Orleans ("Western Water-ways" series)	Follows 1884 Washington, D.C., convention. Delegates from industrial and commercial firms and organizations, plus mayors and governors from states between Minnesota and Louisiana. Chair proclaims this the most important meeting since 1847 Chicago convention.	CIWW New Orleans, *Proceedings;* CIWW Memphis, *Proceedings*, 84

1885	Northwestern Waterways Convention, St. Paul	Nine hundred twenty-six delegates from Midwest seeking federal action to complete the improvement of the Mississippi River and to establish a navigable route from Lake Superior to the Mississippi (including Charles A. Pillsbury, representing the Minneapolis Mills Association).	Northwestern Waterways Convention, St. Paul, *Official Proceedings*, 22
1885	River and Harbor Improvement Convention, Tuscaloosa, Alabama	Nearly 200 delegates from Alabama promoting federal river improvement in the state.	River and Harbor Improvement Convention of 1885 Tuscaloosa, *Memorial*
1887	Convention on the Improvement of Western Water-ways, Memphis ("Western Water-ways series)	The people of the valley are unified in the question of river improvement. Called by Memphis Merchants' Exchange Western Water-ways Executive Committee. About 400 delegates.	CIWW Memphis, *Proceedings*
1887	Mississippi River Convention, Quincy, Illinois ("Western Water-ways" series)	Follows 1885 New Orleans convention. Why is river commerce declining? Seeking improvement of tributary mouths.	CIWW Memphis, *Proceedings*, 90; Mississippi River Convention Quincy, *Proceedings*
1887	Peoria, Illinois	A call to restore the ancient connection between the Great Lakes and the Mississippi.	CIWW Memphis, *Proceedings*, 57

1888	Convention on the Improvement of Western Water-ways, Washington, D.C. ("Western Waterways" series)		CIWW Memphis, *Proceedings*, 90
1888	Savannah Valley Convention, Augusta, Georgia	Local meeting promoting Savannah River improvement by federal government.	Savannah Valley Convention, *Memorial to the Congress of the United States, February 1, 1888*
1889	Convention on the Improvement of the Western Water-Ways, Cincinnati ("Western Waterways" series)	Improvement of Mississippi River and its tributaries is vital to the development of the region.	CIWW Cincinnati, *Proceedings*
1890s	Outlet Convention, Natchez	Supporting artificial outlets as an engineering solution.	Reuss, *Designing the Bayous*, 86
1890s	Levee commissioners convention, Baton Rouge	Opposing outlets as an engineering option and affirming levee building on the main stem of the Mississippi.	Reuss, *Designing the Bayous*, 86
1895	Vicksburg Waterways Convention		Harrison, *Levee Districts and Levee Building in Mississippi*, 134
1896	River Improvement Meeting, Chicago	Meeting of executives from railroads, waterway associations, coal, lumber, and engineering firms, seeking to make the Chicago River a link to Northeast and West.	River Improvement Meeting Chicago 1896, *Chicago!*

1907	Memphis	Held on the eve of the opening of the Panama Canal. More than 10,000 attended, hearing a speech by President Theodore Roosevelt in favor of building facilities for hydroelectric power, irrigation, and flood control.	Barry, *Rising Tide*, 114–15
1927	Chicago Flood Control Conference	Called to create pressure for a permanent program for the Mississippi River, attracting several thousand delegates, including 150 elected representatives.	Barry, *Rising Tide*, 400

Appendix 2

MISSISSIPPI RIVER LEVEE ASSOCIATION EXECUTIVE COMMITTEE

Wm. Wilms	Vice president, Chicago Mill and Lumber Co., Chicago
H. L. Block	President, Un. Sand and Material Co., St. Louis
A. S. Caldwell	Banker, Memphis (MRLA president)
T. O. Vinton	President, Bank of Commerce and Trade Co., Memphis
N. C. Perkins	Vice president, Union and Planters Bank and Trust Co., Memphis
T. K. Riddick	Lawyer, Memphis
L. K. Salsbury	President, Delta Planting Co., Scott, Mississippi
W. H. Russe	President, Russe and Burgess, Inc., Memphis
Paul Dillard	President, Dillard and Coffin Co., Memphis
C. P. J. Mooney	Managing editor, Commercial Appeal, Memphis
O. N. Killough	President, St. Francis Levee Board, Wynne, Arkansas
E. M. Ford	Planter, Deckerville, Arkansas
C. R. Smith	Planter, Cleveland, Mississippi
A. L. Stone	Planter, Dunleith, Mississippi
LeRoy Percy	Lawyer and planter, Greenville, Mississippi
J. W. Cutrer	President, Yazoo-Mississippi Delta Levee Board, Clarksdale, Mississippi
P. M. Harding	President, Delta Trust and Banking Co, Vicksburg, Mississippi
W. H. Fitzhugh	Merchant, Vicksburg, Mississippi
Chas. Scott	Lawyer and planter, Rosedale, Mississippi
H. C. Leake	President, Pontchartrain Levee Board, New Orleans
J. T. McClellan	Planter, Tallulah, Louisiana
Walter Sillers	President, Mississippi Levee District, Rosedale, Mississippi
E. M. Allen	Real estate, Helena, Arkansas

From Mississippi River Levee Association, *Mississippi River Levee Association.*

Notes

Preface

1. *Water for People, Water for Life*, 2003.

2. Reuss, "The Development of American Water Resources"; 52, 55.

3. For example, Arnold, *The Evolution of the 1936 Flood Control Act*; Camillo and Pearcy, *Upon Their Shoulders*; Christie, "The Mississippi Valley Committee"; Frank, *The Development of the Federal Program of Flood Control on the Mississippi River*; Hubbard, *Origins of the* TVA; Leuchtenburg, "Roosevelt, Norris, and the 'Seven Little TVA's.'"

4. Barry, *Rising Tide*; Harrison, *Alluvial Empire*; Harrison, *Levee Districts and Levee Building in Mississippi*; Kelley, *Battling the Inland Sea*; Kelley, *Gold vs. Grain*; Kelley, "Taming the Sacramento"; Kelley, "Forgotten Giant"; Reuss, *Designing the Bayous.*

5. Moore, *Social Origins of Dictatorship and Democracy.*

1 Infrastructure Builds the State

1. Building practices in the United States differ from practices in other countries that share similar legal codes. Until recently, people in Britain have seldom built on floodplains. Britain's central government does little flood control work. The Australian government provides little disaster prevention assistance of any kind. Burton, "Some Aspects of Flood Loss Reduction in England and Wales"; Leivesley, "Natural Disasters in Australia"; Stone, "Social Mobility in England, 1500–1700."

2. Rokkan, *Cities, Elections, Parties*; Sahlins, *Boundaries.*

3. Steinmetz, "Introduction," 9.

4. Other water programs aiding production were the 1902 Reclamation Act and the 1920 Water Power Act. Hays, *Conservation and the Gospel of Efficiency.* The flood control program was the only one of these three that was not expected to pay for itself, although the other two have drawn extensively from general revenues since being created.

5. Tilly, "Epilogue."

6. Weber, "Politics as a Vocation," 78.

7. Mann, *The Sources of Social Power*; Sahlins, *Boundaries.*

8. Collins, "Some Principles of Long-Term Social Change."

9. Tilly, *The Contentious French.*

10. Fox, "New Approaches to International Human Rights."

11. Brenner et al., "Introduction"; Schreurs, "Conservation, Development, and State Sovereignty."

12. Tocqueville, *The Old Regime and the French Revolution.*

13. See, for example, Skocpol, *States and Social Revolutions.*

14. Mitchell, "The Limits of the State."

15. Laumann and Knoke, *The Organizational State*; Miliband, *The State in Capitalist Society*; Mitchell, "The Limits of the State"; Mitchell, "Society, Economy, and the State Effect."

16. Gellner, *Nations and Nationalism*; Mann, *The Sources of Social Power,* Vol. 1; Tilly, *Coercion, Capital, and European States, AD 990–1990*; Tilly, "War Making and State Making as Organized Crime."

17. Schaeffer, *Severed States.*

18. Gellner, *Nations and Nationalism.*

19. Sahlins, *Boundaries.*

20. Rokkan, *Cities, Elections, Parties.*

21. Mitchell, "The Limits of the State"; Mitchell, "Society, Economy, and the State Effect," 95.

22. Sahlins, *Boundaries*; Wilson and Donnan, *Border Identities.*

23. Levi, *An Introduction to Legal Reasoning,* 60.

24. Friedman, *A History of American Law*; Harris, *Origin of the Land Tenure System in the United States*; Llewellyn, *The Bramble Bush.*

25. Harris, *Origin of the Land Tenure System in the United States.*

26. Friedman, *A History of American Law*; Gates and Swenson, *History of Public Land Law Development*; Horwitz, *The Transformation of American Law*; Kulikoff, *The Agrarian Origins of American Capitalism.*

27. Llewellyn, *The Bramble Bush,* 10–11; Polanyi, *The Great Transformation*; Tilly, *As Sociology Meets History*; Tilly, *The Contentious French*; Weber, *Economy and Society.*

28. Moore, *Social Origins of Dictatorship and Democracy.*

29. Harvey, *Justice, Nature and the Geography of Difference*; Lefebvre, *The Production of Space.*

30. Ashton and Philpin, *The Brenner Debate*; Mann, *Agrarian Capitalism in Theory and Practice*; Mooney, "Labor Time and Capitalist Development in Agriculture."

31. Pfeffer, "Social Origins of Three Systems of Farm Production in the United States."

32. Davis, *The Federal Principle.*

33. Ibid.; Reagan and Sanzone, *The New Federalism*; Short, *The Development of National Administrative Organization in the United States.*

34. Gellner, *Nations and Nationalism.*

35. Rubinson, "Political Transformation in Germany and the United States."

36. Pfeffer, "Social Origins of Three Systems of Farm Production in the United States."

37. Sahlins, *Boundaries.*

2 The Founding Principles of River Development

1. Armstrong, Robinson, and Hoy, *History of Public Works in the United States, 1776–1976*; Morison, *The Oxford History of the American People.*

2. Scheiber, "Federalism and the American Economic Order, 1789–1910."

3. Horwitz, *The Transformation of American Law*; Kulikoff, *The Agrarian Origins of American Capitalism*; Weaver, "Beyond the Fatal Shore"; Wilentz, "Society, Politics, and the Market Revolution, 1815–1848."

4. Harris, *Origin Of the Land Tenure System In the United States.*

5. Gates and Swenson, *History of Public Land Law Development*; Lee, *Reclaiming the American West.*

6. Harris, *Origin of the Land Tenure System in the United States.*

7. Dick, *The Lure of the Land*; Gates, "The Nationalizing Influence of the Public Lands."

8. Dick, *The Lure of the Land*; Ferrell and Natkiel, *Atlas of American History*; Robbins, *Our Landed Heritage.*

9. Dobbin, *Forging Industrial Policy*; Rohrbough, *The Land Office Business.*

10. Dick, *The Lure of the Land*; Friedman, *A History of American Law*; Gates and Swenson, *History of Public Land Law Development.*

11. Friedman, *A History of American Law*; Horwitz, *The Transformation of American Law*; Petulla, *American Environmental History*; Scheiber, "Federalism and the American Economic Order, 1789–1910."

12. Moore, *Social Origins of Dictatorship and Democracy.*

13. Gates, "The Nationalizing Influence of the Public Lands."

14. Coulter, "The Efforts of the Democratic Societies of the West to Open the Navigation of the Mississippi"; Gates and Swenson, *History of Public Land Law Development.*

15. Ferrell and Natkiel, *Atlas of American History*; Morison, *The Oxford History of the American People*; Petulla, *American Environmental History*; Wolf, *Europe and the People Without History.*

16. Gates and Swenson, *History of Public Land Law Development*; Morison, *The Oxford History of the American People*; Shallat, *Structures in the Stream.*

17. Morison, *The Oxford History of the American People*; Reichley, *The Life of the Parties.*

18. Gates and Swenson, *History of Public Land Law Development*; Morison, *The*

Oxford History of the American People; Nettels, "The Mississippi Valley and the Constitution"; Robbins, *Our Landed Heritage*; Shallat, *Structures in the Stream.*

19. Minicucci, "The 'Cement of Interest.' "

20. Armstrong, Robinson, and Hoy, *History of Public Works in the United States, 1776–1976*; Dick, *The Lure of the Land*; Nettels, "The Mississippi Valley and the Constitution"; Reichley, *The Life of the Parties*; Shallat, *Structures in the Stream.*

21. Gates, "The Nationalizing Influence of the Public Lands."

22. Nettels, "The Mississippi Valley and the Constitution"; Shallat, *Structures in the Stream.*

23. Morison, *The Oxford History of the American People*; Nettels, "The Mississippi Valley and the Constitution"; Reichley, *The Life of the Parties.*

24. Armstrong, Robinson, and Hoy, *History of Public Works in the United States, 1776–1976*; Nettels, "The Mississippi Valley and the Constitution"; Petulla, *American Environmental History*; Shallat, *Structures in the Stream.*

25. Nettels, "The Mississippi Valley and the Constitution."

26. Shallat, *Structures in the Stream.*

27. Wilentz, "Society, Politics, and the Market Revolution, 1815–1848."

28. Armstrong, Robinson, and Hoy, *History of Public Works in the United States, 1776–1976.*

29. Goodrich, *Government Promotion of American Canals and Railroads, 1800–1890.*

30. Armstrong, Robinson, and Hoy, *History of Public Works in the United States, 1776–1976*; Petulla, *American Environmental History*; Reichley, *The Life of the Parties.*

31. Goodrich, *Government Promotion of American Canals and Railroads, 1800–1890.*

32. Armstrong, Robinson, and Hoy, *History of Public Works in the United States, 1776–1976.*

33. Ibid.; Dobbin, *Forging Industrial Policy*; Petulla, *American Environmental History.*

34. Friedman, *A History of American Law*; Petulla, *American Environmental History*; Scheiber, "Federalism and the American Economic Order, 1789–1910."

35. Armstrong, Robinson, and Hoy, *History of Public Works in the United States, 1776–1976*; Scheiber, "Federalism and the American Economic Order, 1789–1910"; Shallat, *Structures in the Stream.*

36. Wilentz, "Society, Politics, and the Market Revolution, 1815–1848."

37. Ibid.; Wolf, *Europe and the People Without History.*

38. Petulla, *American Environmental History*; Scheiber, "Federalism and the American Economic Order, 1789–1910."

39. Armstrong, Robinson, and Hoy, *History of Public Works in the United States, 1776–1976.*

40. Shallat, *Structures in the Stream.*

41. Scheiber, "Federalism and the American Economic Order, 1789–1910."

42. Kelley, *Battling the Inland Sea*; Shallat, *Structures in the Stream.*

43. Armstrong, Robinson, and Hoy, *History of Public Works in the United States, 1776–1976*; Kelley, *Battling the Inland Sea*; Petulla, *American Environmental History*; Scheiber, "Federalism and the American Economic Order, 1789–1910."

44. Shallat, "Engineering Policy."

45. Childs, *Mighty Mississippi*; Shallat, *Structures in the Stream*; Tompkins, *Riparian Lands of the Mississippi River.*

46. Nettels, "The Mississippi Valley and the Constitution."

47. Shallat, "Engineering Policy"; Shallat, *Structures in the Stream.*

48. Bartlett, *Policy and Power*; Shallat, *Structures in the Stream*; Weigley, *History of the United States Army.*

49. Dobbin, *Forging Industrial Policy*; Mann, *The Sources of Social Power,* Vol. 2; Morison, *The Oxford History of the American People*; Weigley, *History of the United States Army.*

50. Armstrong, Robinson, and Hoy, *History of Public Works in the United States, 1776–1976*; Weigley, *History of the United States Army.*

51. Shallat, "Engineering Policy"; Shallat, *Structures in the Stream*; Weigley, *History of the United States Army.*

52. Maass, *Muddy Waters*; Shallat, "Engineering Policy."

53. Shallat, "Engineering Policy"; Shallat, *Structures in the Stream.*

54. Gates and Swenson, *History of Public Land Law Development*; Shallat, *Structures in the Stream.*

55. Gates and Swenson, *History of Public Land Law Development*; Shallat, "Engineering Policy"; Shallat, *Structures in the Stream*; Wilentz, "Society, Politics, and the Market Revolution, 1815–1848."

56. Gates and Swenson, *History of Public Land Law Development*; Reichley, *The Life of the Parties*; Wilentz, "Society, Politics, and the Market Revolution, 1815–1848."

57. Shallat, *Structures in the Stream*; Wilentz, "Society, Politics, and the Market Revolution, 1815–1848."

58. Gates and Swenson, *History of Public Land Law Development*; Shallat, "Engineering Policy"; Shallat, *Structures in the Stream*; Weigley, *History of the United States Army.*

59. Shallat, "Engineering Policy"; Shallat, *Structures in the Stream.*

60. Ibid.

61. Ibid.

62. Gates and Swenson, *History of Public Land Law Development.*

63. Humphreys and Abbot, *Report Upon the Physics and Hydraulics of the Mississippi River*; Shallat, *Structures in the Stream.*

64. Gates and Swenson, *History of Public Land Law Development*; Reichley,

The Life of the Parties; Wilentz, "Society, Politics, and the Market Revolution, 1815–1848."

65. Congressional Quarterly, *National Party Conventions, 1831–1992*; Gates and Swenson, *History of Public Land Law Development.*

66. Wilentz, "Society, Politics, and the Market Revolution, 1815–1848."

67. Reichley, *The Life of the Parties*; Sitterson, *Sugar Country.*

68. Congressional Quarterly, *National Party Conventions, 1831–1992*; Reichley, *The Life of the Parties.*

69. Armstrong, Robinson, and Hoy, *History of Public Works in the United States, 1776–1976*; Dick, *The Lure of the Land*; Dobbin, *Forging Industrial Policy*; Scheiber, "Federalism and the American Economic Order, 1789–1910"; Shallat, "Engineering Policy"; Shallat, *Structures in the Stream.*

70. Congressional Quarterly, *National Party Conventions, 1831–1992.*

71. Gates, "The Nationalizing Influence of the Public Lands."

Part II

1. Pfeffer, "Social Origins of Three Systems of Farm Production in the United States."

3 The Mississippi River: Becoming the Nation's River

1. U.S. House of Representatives, Committee on Flood Control, *Hearings on Mississippi River Floods.*

2. Northwestern Waterways Convention, St. Paul, *Official Proceedings of the Northwestern Waterways Convention Held at the City of St. Paul, September 3 and 4, 1885.* Hereinafter cited as Northwestern Waterways Convention, *Official Proceedings.*

3. Barry, *Rising Tide.*

4. Ibid.; Elliott, *The Improvement of the Lower Mississippi River for Flood Control and Navigation*; Humphreys and Abbot, *Report Upon the Physics and Hydraulics of the Mississippi River*; McPhee, "Atchafalaya."

5. Shallat, *Structures in the Stream.*

6. Armstrong, Robinson, and Hoy, *History of Public Works in the United States, 1776–1976*; Humphreys and Abbot, *Report Upon the Physics and Hydraulics of the Mississippi River*; Saxon, *Father Mississippi.*

7. McPhee, "Atchafalaya"; Rightor, *Standard History of New Orleans.*

8. Elliott, *The Improvement of the Lower Mississippi River for Flood Control and Navigation.*

9. Morison, *The Oxford History of the American People.*

10. Bartlett, *Policy and Power*; Childs, *Mighty Mississippi*; Gates and Swenson, *History of Public Land Law Development*.

11. Bartlett, *Policy and Power*; Childs, *Mighty Mississippi*.

12. Bartlett, *Policy and Power*.

13. Coulter, "The Efforts of the Democratic Societies of the West to Open the Navigation of the Mississippi."

14. Bartlett, *Policy and Power*; Morison, *The Oxford History of the American People*.

15. Armstrong, Robinson, and Hoy, *History of Public Works in the United States, 1776–1976*; Bartlett, *Policy and Power*; Ferrell and Natkiel, *Atlas of American History*; Morison, *The Oxford History of the American People*.

16. Armstrong, Robinson, and Hoy, *History of Public Works in the United States, 1776–1976*; Elliott, *The Improvement of the Lower Mississippi River for Flood Control and Navigation*; Morison, *The Oxford History of the American People*; Weigley, *History of the United States Army*; Wolf, *Europe and the People Without History*.

17. The Battle of New Orleans made Jackson's political career, even though the battle was fought after the peace treaty had been signed. Morison, *The Oxford History of the American People*.

18. Ferrell and Natkiel, *Atlas of American History*; Wolf, *Europe and the People Without History*.

19. Morison, *The Oxford History of the American People*; Sitterson, *Sugar Country*.

20. Childs, *Mighty Mississippi*.

21. Ferrell and Natkiel, *Atlas of American History*; Sitterson, *Sugar Country*.

22. Dick, *The Lure of the Land*; Gates and Swenson, *History of Public Land Law Development*; Sitterson, *Sugar Country*.

23. Childs, *Mighty Mississippi*; Ferrell and Natkiel, *Atlas of American History*; Winston, "Notes on the Economic History of New Orleans, 1803–1836."

24. Elliott, *The Improvement of the Lower Mississippi River for Flood Control and Navigation*; Harrison, *Levee Districts and Levee Building in Mississippi*.

25. Harrison, *Levee Districts and Levee Building in Mississippi*.

26. McPhee, "Atchafalaya"; Sitterson, *Sugar Country*.

27. Twain, *Life on the Mississippi*, 139–40.

28. Ferrell and Natkiel, *Atlas of American History*; Nettels, "The Mississippi Valley and the Constitution."

29. Childs, *Mighty Mississippi*; Elliott, *The Improvement of the Lower Mississippi River for Flood Control and Navigation*.

30. Shallat, *Structures in the Stream*.

31. Elliott, *The Improvement of the Lower Mississippi River for Flood Control and Navigation*; McPhee, "Atchafalaya"; Nettels, "The Mississippi Valley and the Constitution"; Shallat, "Engineering Policy"; Shallat, *Structures in the Stream*.

32. Shallat, *Structures in the Stream*; Williams, "The Background of the Chicago River and Harbor Convention, 1847."

33. Moore, *Social Origins of Dictatorship and Democracy*; Williams, "The Background of the Chicago River and Harbor Convention, 1847."

34. Childs, *Mighty Mississippi*; Twain, *Life on the Mississippi.*

35. Arnold, *The Evolution of the 1936 Flood Control Act*; Frank, *The Development of the Federal Program of Flood Control on the Mississippi River*; Harrison, *Alluvial Empire*, 60–63, Harrison, *Levee Districts and Levee Building in Mississippi*; Sillers, "Flood Control in Bolivar County, 1838–1924."

36. Harrison, *Levee Districts and Levee Building in Mississippi.*

37. Barry, *Rising Tide*; Elliott, *The Improvement of the Lower Mississippi River for Flood Control and Navigation*; McPhee, "Atchafalaya."

38. Forshey, *Memoir on the Physics of the Mississippi River, and Certain Internal Improvements in the State of Louisiana.*

39. Sitterson, *Sugar Country.*

40. Forshey, *Memoir on the Physics of the Mississippi River, and Certain Internal Improvements in the State of Louisiana.*

41. Shallat, *Structures in the Stream.*

42. Ibid.

43. McPhee, "Atchafalaya."

4 The Mississippi River: Resentment Leading to Civil War

1. Shallat, *Structures in the Stream.*

2. Tilly, *The Contentious French.*

3. Childs, *Mighty Mississippi.*

4. St. Louis Chamber of Commerce, *Proceedings of the St. Louis Chamber of Commerce, in Relation to the Improvement of the Navigation of the Mississippi River and Its Principal Tributaries and the St. Louis Harbor*; St. Louis Citizens, *Memorial of the Citizens of St. Louis, Missouri, to the Congress of the United States.*

5. Cincinnati City Council, *Memorial of the Citizens of Cincinnati, to the Congress of the United States, Relative to the Navigation of the Ohio and Mississippi Rivers.*

6. Harrison, *Levee Districts and Levee Building in Mississippi.*

7. Frank, *The Development of the Federal Program of Flood Control on the Mississippi River.*

8. Harbor and River Convention [of 1847] Chicago, *Memorial to the Congress of the United States of the Executive Committee of the Convention Held at Chicago, July 5, 1847.*

9. Harrison, *Levee Districts and Levee Building in Mississippi.*

10. Williams, "The Background of the Chicago River and Harbor Convention, 1847."

11. Morison, *The Oxford History of the American People.*

12. Shallat, *Structures in the Stream*; Williams, "The Background of the Chicago River and Harbor Convention, 1847"; Williams, "The Chicago River and Harbor Convention, 1847."

13. Shallat, *Structures in the Stream*; Williams, "The Background of the Chicago River and Harbor Convention, 1847."

14. Delegates from the City of St. Louis, *The Commerce and Navigation of the Valley of the Mississippi and Also That Appertaining to the City of St. Louis*; Hall, *Chicago River-and-Harbor Convention July 5th, 6th, and 7th, 1847*; St. Louis Chamber of Commerce, *Proceedings of the St. Louis Chamber of Commerce, in Relation to the Improvement of the Navigation of the Mississippi River and Its Principal Tributaries and the St. Louis Harbor.*

15. Delegates from the City of St. Louis, *The Commerce and Navigation of the Valley of the Mississippi and Also That Appertaining to the City of St. Louis*; Harbor and River Convention [of 1847] Chicago, *Memorial to the Congress of the United States of the Executive Committee of the Convention Held at Chicago, July 5, 1847*; Harbor and River Convention Chicago, *Proceedings of the Harbor and River Convention Held at Chicago, July Fifth, 1847*; Williams, "The Chicago River and Harbor Convention, 1847."

16. Shallat, *Structures in the Stream*; Williams, "The Background of the Chicago River and Harbor Convention, 1847."

17. Hall, *Chicago River-and-Harbor Convention July 5th, 6th, and 7th, 1847.*

18. Williams, "The Chicago River and Harbor Convention, 1847."

19. *Gen. Cass' Letter to the Harbor and River Convention [Chicago, 1847].*

20. Morison, *The Oxford History of the American People.*

21. Shallat, *Structures in the Stream*; Williams, "The Background of the Chicago River and Harbor Convention, 1847."

22. Childs, *Mighty Mississippi*; Rapids Convention [of 1851,] Burlington, *Proceedings of the Rapids Convention, Held at Burlington, Iowa, on the 23rd and 24th of October, 1851.*

23. Shallat, *Structures in the Stream.*

24. Mar. 2, 1849, 9 Stat. 352.

25. Barry, *Rising Tide.*

26. Hibbard, *A History of the Public Land Policies.*

27. Barry, *Rising Tide.*

28. Sept. 28, 1850, 9 Stat. 519; Mar. 12, 1860, 12 Stat. 3; Elliott, *The Improvement of the Lower Mississippi River for Flood Control and Navigation*; Gates and Swenson, *History of Public Land Law Development.*

29. Barry, *Rising Tide.*

30. U.S. Department of Interior, General Land Office, *Annual Reports of the Department of the Interior. Report of the Commissioner of the General Land Office.*

31. Ibid.

32. Gates and Swenson, *History of Public Land Law Development.*

33. McPhee, "Atchafalaya."

34. Gates, "The Homestead Law in an Incongruous Land System"; Gates and Swenson, *History of Public Land Law Development.*

35. Harrison, *Levee Districts and Levee Building in Mississippi.*

36. Gates and Swenson, *History of Public Land Law Development.*

37. McPhee, "Atchafalaya."

38. These states are Alabama, Florida, Illinois, Indiana, Iowa, Michigan, Minnesota, Missouri, Ohio, Oregon, and Wisconsin. States along the upper Mississippi River, for example, were slower to form levee districts, and officials in these states were less active in flood control activism. There, settlement came later because much of their farmland was of low value and because river flooding was less frequent.

39. Elliott, *The Improvement of the Lower Mississippi River for Flood Control and Navigation.*

40. Harrison, *Levee Districts and Levee Building in Mississippi*; Kelley, *Battling the Inland Sea.*

41. Shallat, *Structures in the Stream*; Williams, "The Chicago River and Harbor Convention, 1847."

42. Barry, *Rising Tide*; Humphreys and Abbot, *Report Upon the Physics and Hydraulics of the Mississippi River*; Shallat, *Structures in the Stream.*

43. Abbot, *A Review of the Report Upon the Physics and Hydraulics of the Mississippi River.*

44. Childs, *Mighty Mississippi*; Twain, *Life on the Mississippi*; Williams, "The Chicago River and Harbor Convention, 1847."

45. Weigley, *History of the United States Army.*

46. Childs, *Mighty Mississippi*; Ferrell and Natkiel, *Atlas of American History*; Gates, *Agriculture and the Civil War*; Gates and Swenson, *History of Public Land Law Development*; Shallat, "Engineering Policy"; Twain, *Life on the Mississippi.*

47. Armstrong, Robinson, and Hoy, *History of Public Works in the United States, 1776–1976*; Chandler, *Land Title Origins*; Gates, "The Nationalizing Influence of the Public Lands"; Gates and Swenson, *History of Public Land Law Development*; Hibbard, *A History of the Public Land Policies.*

48. Harrison, *Alluvial Empire*; Harrison, *Levee Districts and Levee Building in Mississippi*; Kelley, *Battling the Inland Sea.*

49. Harrison, *Levee Districts and Levee Building in Mississippi.*

50. H. W. Walter, *De Bow's Commercial Review*, October 1858, 438–39, quoted ibid., 13.

51. Ibid.; Humphreys and Abbot, "Report Upon the Physics and Hydraulics of the Mississippi River."

52. Harrison, *Levee Districts and Levee Building in Mississippi*; Sitterson, *Sugar Country.*

53. Harrison, *Levee Districts and Levee Building in Mississippi.*

54. Armstrong, Robinson, and Hoy, *History of Public Works in the United States, 1776–1976.*

55. Ibid.; Frank, *The Development of the Federal Program of Flood Control on the Mississippi River*; Harrison, *Alluvial Empire*, 62–63, Harrison, *Levee Districts and Levee Building in Mississippi*; U.S. House of Representatives, Committee on Flood Control, *Hearings on Mississippi River Floods.*

56. Ferrell and Natkiel, *Atlas of American History*; Sitterson, *Sugar Country.*

57. Humphreys and Abbot, "Report Upon the Physics and Hydraulics of the Mississippi River."

58. Harrison, *Levee Districts and Levee Building in Mississippi.*

59. Elliott, *The Improvement of the Lower Mississippi River for Flood Control and Navigation*; Humphreys and Abbot, "Report Upon the Physics and Hydraulics of the Mississippi River."

60. Morison, *The Oxford History of the American People.*

61. Shannon, *The Farmer's Last Frontier.*

62. Armstrong, Robinson, and Hoy, *History of Public Works in the United States, 1776–1976*; Ferrell and Natkiel, *Atlas of American History.*

63. Childs, *Mighty Mississippi.*

64. Bensel, *Sectionalism and American Political Development, 1880–1980*; Bensel, *Yankee Leviathan*; Gates, *Agriculture and the Civil War*; Keller, *Affairs of State*; Moore, *Social Origins of Dictatorship and Democracy*; Woodward, *Reunion and Reaction.*

65. Bensel, *Sectionalism and American Political Development*; Bensel, *Yankee Leviathan.*

66. Bensel, *Yankee Leviathan.*

67. Moore, *Social Origins of Dictatorship and Democracy.*

68. Trotsky, *The Russian Revolution*, 201.

5 The Mississippi River: Postwar Reunification, Postwar Aid

1. Armstrong, Robinson, and Hoy, *History of Public Works in the United States, 1776–1976*; Ferrell and Natkiel, *Atlas of American History*; Petulla, *American Environmental History*; Shannon, *The Farmer's Last Frontier.*

2. Dobbin, *Forging Industrial Policy*; Lowi and Harpham, "Political Theory and Public Policy"; Orloff, "The Political Origins of America's Belated Welfare State"; Roy, *Socializing Capital*; Skocpol, *Protecting Soldiers and Mothers*; Skocpol and Ikenberry, "The Political Formation of the American Welfare State in Historical and Comparative Perspective."

3. Skowronek, *Building a New American State.*

4. Bensel, *Yankee Leviathan.*

5. Skocpol, *Protecting Soldiers and Mothers.*

6. Rokkan, *Cities, Elections, Parties.*

7. Moore, *Social Origins of Dictatorship and Democracy.*

8. Kane and Mann, "A Theory of Early Twentieth-Century Agrarian Politics"; Kulikoff, *The Agrarian Origins of American Capitalism.*

9. Moore, *Social Origins of Dictatorship and Democracy*, 149.

10. Bensel, *Yankee Leviathan.*

11. Ibid.; Grantham, *The Life and Death of the Solid South*; Moore, *Social Origins of Dictatorship and Democracy.*

12. Barry, *Rising Tide*; Elliott, *The Improvement of the Lower Mississippi River for Flood Control and Navigation*; Ferrell and Natkiel, *Atlas of American History*; Gates, *Agriculture and the Civil War*; Harrison, *Levee Districts and Levee Building in Mississippi*; Shallat, *Structures in the Stream.*

13. Shallat, *Structures in the Stream*; Tompkins, *Riparian Lands of the Mississippi River.*

14. Frank, *The Development of the Federal Program of Flood Control on the Mississippi River*; Keller, *Affairs of State*; Morison, *The Oxford History of the American People.*

15. Foner, *Reconstruction*; Harrison, *Levee Districts and Levee Building in Mississippi*; Keller, *Affairs of State*; Woodward, *Origins of the New South, 1877–1913*; Woodward, *Reunion and Reaction.*

16. Woolfolk, *The Cotton Regency.*

17. Seip, *The South Returns to Congress*; Shannon, *The Farmer's Last Frontier.*

18. Harrison, *Levee Districts and Levee Building in Mississippi*; Sitterson, *Sugar Country*; Union Merchants' Exchange of St. Louis, *Report of the Committee on Improvement of the Mississippi River and Tributaries*; Wells, *The National Importance of Rebuilding and Repairing the Levees, on the Banks of the Mississippi. . . .*

19. Seip, *The South Returns to Congress.*

20. Aiken, "The Decline of Sharecropping in the Lower Mississippi River Valley"; Foner, *Reconstruction*; Mann, *Agrarian Capitalism in Theory and Practice.*

21. Barry, *Rising Tide*; Gates, *Agriculture and the Civil War.*

22. Union Merchants' Exchange of St. Louis, *Report of the Committee on Improvement of the Mississippi River and Tributaries.*

23. Foner, *Reconstruction*; Wells, *The National Importance of Rebuilding and Repairing the Levees, on the Banks of the Mississippi. . . .*

24. Harrison, *Levee Districts and Levee Building in Mississippi*; Sitterson, *Sugar Country.*

25. E. O. Stanard, president of the Union Merchant's Exchange of St. Louis in the 1860s and governor of Missouri in the 1870s, appears to have been the busiest of activists at mid-century, acting as convention president, convention organizer and vice president, executive committee member convention president, and delegate. Convention on the Improvement of the Western Wa-

terways Memphis, *Proceedings of the Convention on the Improvement of the Western Water-ways, Together With a Brief Memorial to Congress* (hereafter cited as CIWW Memphis, *Proceedings*); Convention for the Improvement of the Mississippi River Washington, D.C., *Proceedings of the Convention for the Improvement of the Mississippi River and Its Navigable Tributaries, Together With a Brief Memorial to Congress, Held at Washington, D.C., Feb. 5, 6, 7, 1884* (hereafter cited as CIWW Washington, *Proceedings*); Mississippi River Improvement Convention, *A Memorial to Congress to Secure an Adequate Appropriation for a Prompt and Thorough Improvement of the Mississippi River* (hereafter cited as MRIC, *Memorial*); Mississippi River Improvement Convention, *Official Report of the Proceedings of the Mississippi River Improvement Convention* (hereafter cited as MRIC, *Official Report*); Mississippi River Improvement Convention Dubuque 1866, *Proceedings of the Mississippi River Improvement Convention Held at Dubuque, Iowa, February 14th and 15th, 1866* (hereafter cited as MRIC Dubuque, *Proceedings*); Northwestern Waterways Convention, *Official Proceedings*; River Improvement Convention Quincy, *Memorial of the Quincy River Improvement Convention, October 15, 1879 to the Congress of the United States, in Regard to the Improvement of the Mississippi River and its Tributaries* (hereafter cited as River Improvement Convention Quincy, *Memorial*); River Improvement Convention St. Louis, *Proceedings of the River Improvement Convention, Held in St. Louis, February 12 and 13, 1867* (hereafter cited as River Improvement Convention St. Louis, *Proceedings*).

Caleb G. Forshey, a professor of civil engineering at Jefferson College, was a field researcher for Humphreys for the army's study published in 1861, a consultant to the Louisiana Legislature's commission on levees, an executive committee member, and a speaker who presented plans for a southern rail route to the Pacific at the 1845 river convention in Memphis. Barry, *Rising Tide*; Forshey, *Memoir on the Physics of the Mississippi River, and Certain Internal Improvements in the State of Louisiana*; MRIC, *Memorial*; Williams, "The Background of the Chicago River and Harbor Convention, 1847."

Captain James B. Eads was an independent engineer who built ironclad gunboats for the Union during the Civil War and became a hero to many in the South for clearing the mouth of the Mississippi River. Eads acted as an executive committee member, wrote letters of support, gave a speech as representative and active member of the Union Merchants' Exchange of St. Louis, and wrote many reports to Congress in the course of getting contracts for bridge, jetty, and other river projects, including a review of the work of the commission of engineers ordered by Congress in 1874. Barry, *Rising Tide*; Childs, *Mighty Mississippi*; Congressional Convention St. Louis, *Proceedings of the Congressional Convention Held in the City of St. Louis on 13th, 14th and 15th days of May 1873* (hereafter cited as Congressional Convention St. Louis, *Proceedings*); Eads, *Physics and Hydraulics of the Mississippi River*; Harrison, *Levee Districts and Levee Building in Mississippi*; River and Harbor Improvement Convention of 1885 Tuscaloosa, *Memorial*; River Improvement Convention St. Louis, *Proceedings.*

Henry C. Haarstick was president of the St. Louis and Mississippi Valley Transportation Co. and vice president of the St. Louis Commercial Club, resolutions committee member, executive committee member, and delegate. CIWW Cincinnati, *Proceedings of the Convention on the Improvement of the Western Water-Ways, Held at Cincinnati, September 4th and 5th, 1889* (hereafter cited as CIWW Cincinnati, *Proceedings*); CIWW Memphis, *Proceedings*; Commercial Club of St. Louis, *Hon. W. M. Fishback (Governor of Arkansas); Hon. W. C. Stone (Governor of Missouri), and Mr. H. C. Haarstick (President of the St. Louis and Mississippi Valley Transportation Co)*; Union Merchants' Exchange of St. Louis, *Memorial of the Union Merchants' Exchange of St. Louis to the Forty-third Congress of the United States.*

J. H. Murphy, E. H. Thayer, S. M. Clark, and E. M. Dickey (of Iowa), Charles E. Cox and Alton municipal worker George R. Hewitt (of Illinois), W. F. Phelps (of Minnesota), John T. Teele and St. Louis city government employee Frank L. Johnson (of Missouri), Frank Gaiennie and Nathan Cole (of the St. Louis Merchants' Exchange), E. H. Allen and T. B. Ballene (of the Kansas City, Missouri, Board of Trade) attended conventions in Memphis and St. Paul. CIWW Memphis, *Proceedings*; Northwestern Waterways Convention, *Official Proceedings.*

Other activists include Joseph Brown, mayor of St. Louis and later congressional representative; J. W. Clapp, on the executive committee of a Memphis convention and delegate to the 1887 Peoria convention; Sam M. Clark and O. C. Merriman, on the board of a Davenport meeting; John H. Gear, former governor of Iowa; congressional representative Col. William Hatch, who chaired a Quincy convention; M. Y. Johnson, congressional representative from Illinois; Jesse McIntyre, of the Red Wing Board of Trade; John A. Scudder, president of the Union Merchants' Exchange of St. Louis and executive committee member of two conventions; and Professor S. Waterhouse of St. Louis.

Brown: Congressional Convention St. Louis, *Proceedings*; MRIC, *Memorial.*

Clapp: CIWW Memphis, *Proceedings.*

Clark and Merriman: Northwestern Waterways Convention, *Official Proceedings*; River and Canal Improvement Convention Davenport, *Proceedings of the River and Canal Improvement Convention, Davenport, Iowa, 1881* (hereafter cited as River and Canal Improvement Convention Davenport, *Proceedings*).

Gear: CIWW Memphis, *Proceedings*; Mississippi River Convention Quincy, *Proceedings of the Mississippi River Convention Held in the Rooms of the Young Men's Business Association, Quincy, Ill* [*sic.*], *Thursday, October 13th, 1887* (hereafter cited as Mississippi River Convention Quincy, *Proceedings*); Northwestern Waterways Convention, *Official Proceedings.*

Hatch: Mississippi River Convention Quincy, *Proceedings*; Northwestern Waterways Convention, *Official Proceedings.*

Johnson: MRIC, *Memorial*; Northwestern Waterways Convention, *Official Proceedings*.

McIntyre: Northwestern Waterways Convention, *Official Proceedings*; River Improvement Convention St. Louis, *Proceedings*.

Scudder: CIWW Memphis, *Proceedings*; MRIC, *Memorial*.

Waterhouse: MRIC, *Memorial*; River Improvement Convention St. Louis, *Proceedings*.

26. Camillo and Pearcy, *Upon Their Shoulders*; Seip, *The South Returns to Congress*; Sitterson, *Sugar Country*.

27. Camillo and Pearcy, *Upon Their Shoulders*; Eads, *Physics and Hydraulics of the Mississippi River*; Hilgard, *Reclamation of the Alluvial Basin of the Mississippi River*; Seip, *The South Returns to Congress*; Shallat, *Structures in the Stream*; Tompkins, *Riparian Lands of the Mississippi River*; U.S. War Department, Commission of Engineers, *Report of the Commission of Engineers Appointed to Investigate and Report a Permanent Plan for the Reclamation of the Alluvial Basin of the Mississippi River Subject to Inundation*.

28. Foner, *Reconstruction*; Grantham, *The Life and Death of the Solid South*; Woodward, *Origins of the New South, 1877–1913*; Woodward, *Reunion and Reaction*.

29. Grantham, *The Life and Death of the Solid South*.

30. Foner, *Reconstruction*; Woodward, *Origins of the New South, 1877–1913*; Woodward, *Reunion and Reaction*.

31. Foner, *Reconstruction*.

32. Ibid.; Seip, *The South Returns to Congress*; Woodward, *Origins of the New South, 1877–1913*; Woodward, *Reunion and Reaction*.

33. Woodward, *Reunion and Reaction*.

34. Grantham, *The Life and Death of the Solid South*.

35. Ibid.; Woodward, *Reunion and Reaction*.

36. Grantham, *The Life and Death of the Solid South*; Moore, *Social Origins of Dictatorship and Democracy*; Woodward, *Origins of the New South, 1877–1913*.

37. Camillo and Pearcy, *Upon Their Shoulders*, ch. 3; Reuss, "Andrew A. Humphreys and the Development of Hydraulic Engineering."

38. Camillo and Pearcy, *Upon Their Shoulders*.; June 28, 1879, c. 43, 21 Stat., 37–38; Tompkins, *Riparian Lands of the Mississippi River*.

39. Frank, *The Development of the Federal Program of Flood Control on the Mississippi River*; Sitterson, *Sugar Country*.

40. Camillo and Pearcy provide details about congressional rules fights and engineering debates that led the Mississippi River Commission to this approach. Camillo and Pearcy, *Upon Their Shoulders*.

41. Ferguson, *History of the Improvement of the Lower Mississippi River for Flood Control and Navigation, 1932–1939*; U.S. Congress, *Congressional Record*.

42. Harrison, *Levee Districts and Levee Building in Mississippi*; Keller, *Affairs of State*.

43. Skowronek, *Building a New American State.*

44. Seip, *The South Returns to Congress.*

6 The Sacramento River: Miners versus Farmers

1. Rokkan, *Cities, Elections, Parties.*

2. Johnson, Haslam, and Dawson, *The Great Central Valley*; Kelley, *Battling the Inland Sea*; U.S. Army Corps of Engineers, Sacramento District, *Sacramento River Aerial Atlas.*

3. Johnson, Haslam, and Dawson, *The Great Central Valley*; Kelley, *Battling the Inland Sea.*

4. Thompson, *The Settlement Geography of the Sacramento-San Joaquin Delta, California.*

5. Bancroft, *History of California, 1860–1890*; Beck and Haase, *Historical Atlas of California*; Thompson, *The Settlement Geography of the Sacramento-San Joaquin Delta, California.*

6. Ferrell and Natkiel, *Atlas of American History.*

7. Morison, *The Oxford History of the American People.*

8. Ferrell and Natkiel, *Atlas of American History*; Gates and Swenson, *History of Public Land Law Development.*

9. Morison, *The Oxford History of the American People.*

10. Rohe, "Man and the Land: Mining's Impact in the Far West."

11. Bancroft, *History of California, 1860–1890*; Caughey and Hundley, *California.*

12. Gates and Swenson, *History of Public Land Law Development*; Shinn, *Mining Camps.*

13. Chandler, *Land Title Origins*; Gates and Swenson, *History of Public Land Law Development.*

14. Beck and Haase, *Historical Atlas of California*; May, *Origins of Hydraulic Mining in California.*

15. Liebman, *California Farmland.*

16. Caughey and Hundley, *California*; Johnson, Haslam, and Dawson, *The Great Central Valley.*

17. Brienes, "Sacramento Defies the Rivers, 1858–1878"; Caughey and Hundley, *California*; Gates and Swenson, *History of Public Land Law Development*; Shinn, *Mining Camps.*

18. State of California, Swamp Land Commissioners, *First Annual Report of the Swamp Land Commissioners*; Thompson, *The Settlement Geography of the Sacramento-San Joaquin Delta, California*; U.S. House of Representatives, Committee on the Public Lands, *Swamp and Overflowed Lands.*

19. Downey, *Annual Message of Governor John G. Downey.*

20. Bishofberger, "Early Flood Control in the Central Valley"; California Sur-

veyor General, *Annual Report of the Surveyor General of California for the Year 1860*; Liebman, *California Farmland*; Peterson, "The Failure to Reclaim"; Thompson, *The Settlement Geography of the Sacramento-San Joaquin Delta, California.*

21. Brienes, "Sacramento Defies the Rivers, 1858–1878."

22. Kelley, *Battling the Inland Sea.*

23. Bishofberger, "Early Flood Control in the Central Valley"; Brienes, "Sacramento Defies the Rivers, 1858–1878."

24. Kelley, *Gold vs. Grain*; Mann, *After the Gold Rush.*

25. Bean, *Bean's History and Directory of Nevada County*; Mann, *After the Gold Rush.*

26. May, *Origins of Hydraulic Mining in California.*

27. Hagwood, *From North Bloomfield to North Fork*; Jewell, *The Development of Governmental Restraints on Hydraulic Mining in California*; McPhee, "Annals of the Former World: Assembling California—1"; McPhee, "Annals of the Former World: Assembling California—2."

28. Kelley, *Gold vs. Grain*; Wells, *History of Nevada County, California.*

29. Rohe, "Man as Geomorphic Agent: Hydraulic Mining in the American West."

30. Kinyon, *The Northern Mines.*

31. U.S. War Department, Board of Engineers [Biggs Commission], *Mining Debris, California.*

32. Kelley, *Gold vs. Grain.*

33. Ibid.

34. Brewer, *Up and Down California.*

35. Mann, *After the Gold Rush*; Shinn, *Mining Camps.*

36. *Mining and Scientific Press*, 6 January 1877, 9, 12.

37. Hagwood, *Commitment to Excellence.*

38. Thompson, *The Settlement Geography of the Sacramento-San Joaquin Delta, California.*

39. Hagwood, *Commitment to Excellence*; Kahrl, *The California Water Atlas.*

40. Johnson, Haslam, and Dawson, *The Great Central Valley*; Liebman, *California Farmland*; Paul, "The Great California Grain War"; Pisani, *From the Family Farm to Agribusiness.*

41. Kelley, *Battling the Inland Sea*; Liebman, *California Farmland.*

42. Great portions of the Tulare Lake area in the southern San Joaquin Valley were distributed under the Swamp Land Grant Act program. This area was planted in bonanza wheat farms at a later period (from the 1890s through the 1910s), because extensive flooding made the area more difficult to manage. Owners later formed reclamation districts in Tulare. Agricultural users eventually diverted the entire flow of the Sierra rivers that historically fed this extensive lake. The area is now completely converted into farmland (although it does frequently flood from rainwater). This area was subject to state and federal flood control and reclamation bills, but landowners there were less active in getting

such legislation passed than were the landowners in the Sacramento Valley. Liebman, *California Farmland*; Pisani, *From the Family Farm to Agribusiness.*

43. Bishofberger, "Early Flood Control in the Central Valley."

44. Thompson, *The Settlement Geography of the Sacramento-San Joaquin Delta, California.*

45. Liebman, *California Farmland*; Peterson, "The Failure to Reclaim."

46. Mann, *After the Gold Rush*; Thompson and Dutra, *The Tule Breakers.*

47. Kelley, *Gold vs. Grain*; Mann, *After the Gold Rush.*

48. Kelley, *Gold vs. Grain.*

49. Mann, *After the Gold Rush.*

50. Caughey and Hundley, *California*; Pisani, *From the Family Farm to Agribusiness.*

51. Ellis, *Memories*; Kelley, "Taming the Sacramento."

52. Kelley, *Gold vs. Grain.*

7 The Sacramento River: Capitalists Unify for Development

1. Kelley, *Battling the Inland Sea*; Peterson, "The Failure to Reclaim."

2. Kelley, *Battling the Inland Sea.*

3. 352 Stat. Cal. 1861, 355.

4. Kelley, *Battling the Inland Sea*; Peterson, "The Failure to Reclaim."

5. Bishofberger, "Early Flood Control in the Central Valley"; Peterson, "The Failure to Reclaim"; Thompson, *The Settlement Geography of the Sacramento-San Joaquin Delta, California.*

6. Bishofberger, "Early Flood Control in the Central Valley"; Peterson, "The Failure to Reclaim."

7. California State Controller, *Annual Report of the Controller of the State for the 15th and 16th Fiscal Years, 1864 and 1865*; California Surveyor General, *Annual Report of the Surveyor General of California for the Year 1862*; Thompson, *The Settlement Geography of the Sacramento-San Joaquin Delta, California.*

8. Kelley, *Battling the Inland Sea*; Peterson, "The Failure to Reclaim."

9. Bishofberger, "Early Flood Control in the Central Valley"; Brienes, "Sacramento Defies the Rivers, 1858–1878"; Kelley, *Battling the Inland Sea*; Peterson, "The Failure to Reclaim."

10. Commonwealth Club, *Transactions of the Commonwealth Club of California*; Kelley, *Battling the Inland Sea*; Peterson, "The Failure to Reclaim."

11. 570 Stat. Cal. 1871–1872, 835; California Surveyor General, *Biennial Report of the Surveyor General of the State of California, From August 1, 1873 to August 1, 1875*; Commonwealth Club, *Transactions of the Commonwealth Club of California*; Kelley, *Battling the Inland Sea*; Thompson, *The Settlement Geography of the Sacramento-San Joaquin Delta, California.*

12. California Surveyor General, *Statistical Report of the Surveyor General of California for the Years 1869, 1870, and 1871*; Robinson, *Land in California.*

13. Bishofberger, "Early Flood Control in the Central Valley"; Kelley, *Battling the Inland Sea*; Liebman, *California Farmland*; Thompson, *The Settlement Geography of the Sacramento-San Joaquin Delta, California.*

14. Kelley, *Gold vs. Grain*; Kelley, "The Mining Debris Controversy in the Sacramento Valley"; *Mining and Scientific Press*, 6 January 1877, 9, 12.

15. Hagwood, *The California Debris Commission*; Kelley, *Battling the Inland Sea*; Kelley, *Gold vs. Grain*; Kelley, "The Mining Debris Controversy in the Sacramento Valley"; *Mining and Scientific Press*, 6 January 1877, 9, 12.

16. Hagwood, *The California Debris Commission*; Kelley, *Gold vs. Grain*; Kelley, "The Mining Debris Controversy in the Sacramento Valley."

17. *Mining and Scientific Press*, 6 January 1877, 9, 12.

18. Hagwood, *The California Debris Commission*; Kelley, *Battling the Inland Sea*; Kelley, *Gold vs. Grain*; Kelley, "The Mining Debris Controversy in the Sacramento Valley"; Mining and Scientific Press, 7 September 1878, 152.

19. *Mining and Scientific Press*, 12 November 1881, 320–21.

20. *Mining and Scientific Press*, 12 November 1881.

21. *Mining and Scientific Press*, 17 November 1881, 313.

22. Dana, *The Sacramento*; Shinn, *Mining Camps.*

23. *Mining and Scientific Press*, 6 January 1877, 9, 12.

24. Ibid.

25. *Mining and Scientific Press*, 17 November 1881, 313.

26. Caughey and Hundley, *California*; Moorhead, "Sectionalism in the California Constitution of 1879."

27. Kelley, *Battling the Inland Sea.*

28. Moorhead, "Sectionalism in the California Constitution of 1879."

29. Pisani, *From the Family Farm to Agribusiness.*

30. Kelley, *Battling the Inland Sea.*

31. *Mining and Scientific Press*, 6 January 1877, 9, 12.

32. Pisani, *From the Family Farm to Agribusiness.*

33. Kelley, *Battling the Inland Sea.*

34. 429 Stat. Cal. 1877–1878, 634, Hall, *Sacramento Valley River Improvement*; Kelley, *Battling the Inland Sea*; Sacramento River Drainage District, *Report of the Board of Commissioners of the Sacramento River Drainage District to the Governor of California*; Thompson, *The Settlement Geography of the Sacramento-San Joaquin Delta, California.*

35. Hagwood, *The California Debris Commission*; Kelley, *Battling the Inland Sea.*

36. Sacramento River Drainage District, *Report of the Board of Commissioners of the Sacramento River Drainage District to the Governor of California*; Thompson, *The Settlement Geography of the Sacramento-San Joaquin Delta, California.*

37. Hagwood, *The California Debris Commission*; Hall, *Sacramento Valley River Improvement*; Kelley, *Battling the Inland Sea*; Sacramento River Drainage Dis-

trict, *Report of the Board of Commissioners of the Sacramento River Drainage District to the Governor of California.*

38. Hagwood, *The California Debris Commission*; Kelley, *Battling the Inland Sea*; Williams, *The Water and the Power.*

39. Kelley, "Taming the Sacramento."

40. Kelley, *Battling the Inland Sea.*

41. Hagwood, *Commitment to Excellence.*

42. 14 June 1880 c. 211, 21 Stat. 180–197; Kelley, *Battling the Inland Sea.*

43. Elliott, *The Improvement of the Lower Mississippi River for Flood Control and Navigation*; Hagwood, *Commitment to Excellence*; Kelley, *Battling the Inland Sea.*

44. Hagwood, *The California Debris Commission*; Hagwood, *From North Bloomfield to North Fork*; Kelley, *Battling the Inland Sea*; Kelley, *Gold vs. Grain.*

45. Hagwood, *The California Debris Commission.*

46. *Mining and Scientific Press*, 12 November 1881, 320.

47. Quoted in Hagwood, *The California Debris Commission*, 25.

48. Ellis, *Memories*; Hagwood, *The California Debris Commission*; Hagwood, *From North Bloomfield to North Fork.*

49. *Edwards Woodruff v. North Bloomfield Gravel Mining Co., et al.*, 9 Sawy. 441; Hagwood, *The California Debris Commission.*

50. Kelley, *Battling the Inland Sea.*

51. Rohe, "Man as Geomorphic Agent."

52. Ibid.; U.S. Army Corps of Engineers, Chief of Engineers, *Annual Report 1896 Upon the Improvement of San Joaquin and Sacramento Rivers and Tributaries, and of Rivers and Harbors in California North of San Francisco*, 3223.

53. Quoted in Rohe, "Man as Geomorphic Agent."

54. Hagwood, *Commitment to Excellence*; Jewell, *The Development of Governmental Restraints on Hydraulic Mining in California*; Kelley, *Battling the Inland Sea.*

55. Jewell, *The Development of Governmental Restraints on Hydraulic Mining in California.* Mining and Scientific Press, 6 May 1893, 274; U.S. War Department, Board of Engineers [Biggs Commission], *Mining Debris, California.*

56. *Mining and Scientific Press*, 28 November 1891, 350.

57. *Mining and Scientific Press*, 12 November 1881, 320–21.

58. *Mining and Scientific Press*, 5 December 1891, 368.

59. Jewell, *The Development of Governmental Restraints on Hydraulic Mining in California.*

60. *Mining and Scientific Press*, 5 December 1891, 363.

61. *Mining and Scientific Press*, 6 May 1893, 275.

62. Hagwood, *Commitment to Excellence*; Kelley, *Battling the Inland Sea*, Mining and Scientific Press, 6 May 1893, 274, 275.

63. *Mining and Scientific Press*, 14 January 1893, 18.

64. Hagwood, *Commitment to Excellence*; Jewell, *The Development of Governmental Restraints on Hydraulic Mining in California*; Kelley, *Battling the Inland Sea*; *Mining and Scientific Press*, 28 January 1893, 50–51; 6 May 1893, 274, 275.

65. 1 March 1893, c. 183, 27 Stat. 507–11.

66. Kelley, *Battling the Inland Sea; Mining and Scientific Press*, 6 May 1893, 275.

67. Jewell, *The Development of Governmental Restraints on Hydraulic Mining in California; Mining and Scientific Press*, 9 September 1983, 162.

68. Kelley, *Battling the Inland Sea.*

69. Federal orders limiting Chinese (1882) and Japanese (1907) immigration shifted tenancy to other groups. Liebman, *California Farmland.*

70. Kelley, *Battling the Inland Sea*; Thompson, *The Settlement Geography of the Sacramento-San Joaquin Delta, California.*

71. U.S. Army Corps of Engineers, Chief of Engineers, *Annual Report 1896 Upon the Improvement of San Joaquin and Sacramento Rivers and Tributaries, and of Rivers and Harbors in California North of San Francisco.*

72. Kelley, *Battling the Inland Sea*; U.S. Army Corps of Engineers, California Debris Commission, *Enacts the 1902 Rivers and Harbors Act Plan for Yuba River.*

73. *Mining and Scientific Press*, 6 May 1893, 275.

74. Quoted in *Mining and Scientific Press*, 23 September 1893, 194.

75. Hagwood, *From North Bloomfield to North Fork*; Jewell, *The Development of Governmental Restraints on Hydraulic Mining in California*; Rohe, "Man as Geomorphic Agent."

76. Kelley, *Battling the Inland Sea.*

77. Rokkan, *Cities, Elections, Parties.*

8 Federal Aid for the Mississippi and Sacramento Rivers

1. Goodwyn, *The Populist Moment.*

2. Bliss, *Letter Regarding the Rivers and Harbors Bill, October 2, 1882*; Camillo and Pearcy, *Upon Their Shoulders*; National Rivers and Harbors Congress, *Proceedings of the Convention, National Rivers and Harbors Congress*, 1910, 18–24; Swain, *Federal Conservation Policy, 1921–1933.*

3. I will describe in detail only the organizations prominent in flood control activism. Many other national and regional waterway organizations lobbied primarily or solely for navigation improvements.

4. Frank, *The Development of the Federal Program of Flood Control on the Mississippi River.*

5. Barry, *Rising Tide*; Frank, *The Development of the Federal Program of Flood Control on the Mississippi River*; Mississippi River Improvement and Levee Convention, "Resolutions of the Mississippi River Improvement and Levee Convention, Held at Vicksburg, Mississippi, April 30 and May 1, 1890, and Resolution Adopted by the Southern Press Association" (hereafter cited as MRILC, "Resolutions"); Tompkins, *Riparian Lands of the Mississippi River.*

6. John M. Parker served as a board member of the MRILA, member of the executive committee of the Lakes-to-the-Gulf Deep Waterway Association, president of the New Orleans Cotton Exchange and of the New Orleans Board of Trade, director of the Illinois Central Railroad, and governor of Louisiana. Barry, *Rising Tide*; Lakes-to-the-Gulf Deep Waterways Association, *The Deep Waterway Between the Great Lakes and the Gulf of Mexico*; Tompkins, *Riparian Lands of the Mississippi River.*

B. D. Wood linked earlier convention-based activism with organization-based activism, serving as president of the Western Waterways Association, the executive committee that organized the annual "Western Waterways" conventions and as an officer of the MRILA. Tompkins, *Riparian Lands of the Mississippi River.* William Starling was a committee member at the 1887 and 1889 "Western Waterways" conventions, the chief engineer of the Lower Yazoo Levee District, and wrote two articles about levees, one of which appeared in the MRILA publication prepared for Congress. Camillo and Pearcy, *Upon Their Shoulders*; CIWW Memphis, *Proceedings*; William Starling, *The Floods of the Mississippi River: Including an Account of their Principal Causes and Effects: And a Description of the Levee System and Other Means Proposed and Tried for the Control of the River, With a Particular Account of the Great Flood of 1897*; Tompkins, *Riparian Lands of the Mississippi River.*

W. A. Percy, brother of the founder of Greenville, Mississippi, and father of LeRoy Percy, had convinced the governor and legislature to create a new levee board in the Yazoo-Mississippi Delta while a Liquidating Board handled the debts of pre–Civil War levee districts. W. A. Percy controlled that board, built the first railroad line into the Delta, and organized finances for both of these ventures through his local bank and through connections in New York and London. Barry, *Rising Tide.* At the 1887 "Western Waterways" convention, W. A. Percy was a state committee member and gave a speech about levees. CIWW Memphis, *Proceedings.* A publication by the MRILA described him as someone "prominent in levee matters." Tompkins, *Riparian Lands of the Mississippi River,* 10. His son LeRoy, discussed below, succeeded him in levee politics and in control of the plantation. Barry, *Rising Tide.*

7. Woodward, *Origins of the New South, 1877–1913.*

8. Harrison, *Levee Districts and Levee Building in Mississippi*; MRILC, "Resolutions."

9. Merchants' Exchange of St. Louis Mississippi River Improvement Committee, *The Improvement of the Mississippi River*; MRILC, "Resolutions."

10. Quoted in Harrison, *Levee Districts and Levee Building in Mississippi.*

11. Quoted ibid., 119, note 307.

12. MRILC, "Resolutions."

13. Frank, *The Development of the Federal Program of Flood Control on the Mississippi River*; MRILC, "Resolutions"; Tompkins, *Riparian Lands of the Mississippi River.*

14. Marcus, "Environmental Policies in the United States."

15. Quoted in Harrison, *Levee Districts and Levee Building in Mississippi*, 120.

16. Tompkins, *Riparian Lands of the Mississippi River*, 5.

17. Camillo and Pearcy, *Upon Their Shoulders*; Harrison, *Levee Districts and Levee Building in Mississippi*; Starling, *The Floods of the Mississippi River*; Tompkins, *Riparian Lands of the Mississippi River*, 5.

18. Tompkins, *Riparian Lands of the Mississippi River*, 5.

19. Camillo and Pearcy, *Upon Their Shoulders*; Humphreys, *Floods and Levees of the Mississippi River*.

20. Pisani, *From the Family Farm to Agribusiness*.

21. Reclamation and Drainage Association, [report], 8.

22. River Improvement and Drainage Association of California, *Report of the Committee of Twenty-four of the River Improvement and Drainage Association of California* (hereafter cited as RIDA, *Report*).

23. Reclamation and Drainage Association, [report].

24. RIDA, *Report*.

25. Kelley, *Battling the Inland Sea*.

26. Commonwealth Club, *Transactions of the Commonwealth Club of California*.

27. Ibid.; Kelley, *Battling the Inland Sea*.

28. California Commissioner of Public Works, *Report of the Commissioner of Public Works to the Governor of California, Together with the Report of the Commissioner of Public Works Upon the Rectification of the Sacramento and San Joaquin Rivers and Their Principal Tributaries, and the Reclamation of the Overflowed Lands Adjacent Thereto*; Commonwealth Club, *Transactions of the Commonwealth Club of California*; Kelley, *Battling the Inland Sea*; Mills, "The Hydrography of the Sacramento Valley"; River Improvement and Drainage Association of California, *Bulletin*, no. 1, 1904; River Improvement and Drainage Association of California, *Bulletin*, no. 7, 1908, 3.

29. River Improvement and Drainage Association of California, *Bulletin*, no. 1, 1904, 2–4.

30. Mills, "The Hydrography of the Sacramento Valley."

31. "Convention Assembles to Devise Ways and Means to Prevent Recurrence of the Destructive Overflows," 9; River Improvement and Drainage Association of California, *Bulletin*, no. 1, 1904, 4; Sacramento Valley Development Association, *Five Years' Review*.

32. Hall, *Sacramento Valley River Improvement*; River Improvement and Drainage Association of California, *Bulletin*, no. 1, 1904, 6.

33. California Commissioner of Public Works, *Report*; Kelley, *Battling the Inland Sea*; River Improvement and Drainage Association of California, *Bulletin*, no. 4, 1905.

34. Goldman, *Rendezvous With Destiny*; Reichley, *The Life of the Parties*; Wiebe, *The Search for Order, 1877–1920*.

35. Goldman, *Rendezvous With Destiny*; Goodwyn, *The Populist Moment*; Shannon, *The Farmer's Last Frontier*.

36. Burns, *Roosevelt*; Reichley, *The Life of the Parties*.

37. Frank, *The Development of the Federal Program of Flood Control on the Mississippi River*.

38. Hays, *Conservation and the Gospel of Efficiency*; Reichley, *The Life of the Parties*; Rodgers, "In Search of Progressivism."

39. Morison, *The Oxford History of the American People*, vol. 3, 132; Petulla, *American Environmental History*; Robinson, *Water for the West*.

40. Burnham, *Critical Elections and the Mainsprings of American Politics*.

41. De Roos and Maass, "The Lobby That Can't Be Licked."

42. Haupt, "The Mississippi River Problem"; Riche, *The Further Improvement of Our Inland Waterways*.

43. National Rivers and Harbors Congress, *Proceedings of the Convention*, 1914, 18; Swain, *Federal Conservation Policy, 1921–1933*.

44. National Rivers and Harbors Congress, *Proceedings of the Convention*, 1908, 203–5.

45. Arnold, *The Evolution of the 1936 Flood Control Act*; Swain, *Federal Conservation Policy, 1921–1933*.

46. Hart, "Crisis, Community, and Consent in Water Politics"; Maass, *Muddy Waters*.

47. National Rivers and Harbors Congress, *Proceedings of the Convention*, 1906, 16–29.

48. National Rivers and Harbors Congress, *Proceedings of the Convention*, 1906, 93, 136; 1908, 39; 1909, 39, 177; 1911.

49. National Rivers and Harbors Congress, *Proceedings of the Convention*, 1907, 246.

50. Fox, "Economic Value of Improved Waterways."

51. National Rivers and Harbors Congress, *Proceedings of the Convention*, 1909, 177–178; 1914, 25–26.

52. Kelley, *Battling the Inland Sea*.

53. River Improvement and Drainage Association of California, *Bulletin*, no. 3, 1905, 9–12; River Improvement and Drainage Association of California, *Bulletin*, no. 4, 1905; Thompson, *The Settlement Geography of the Sacramento-San Joaquin Delta, California*.

54. Hall, *Sacramento Valley River Improvement*, 6, 13.

55. Kelley, *Battling the Inland Sea*; U.S. Army Corps of Engineers, California Debris Commission, *Enacts the 1902 Rivers and Harbors Act Plan for Yuba River*.

56. U.S. Army Corps of Engineers, California Debris Commission, *Enacts the 1902 Rivers and Harbors Act Plan for Yuba River*.

57. Kelley, *Battling the Inland Sea*.

58. Ibid.; Kelley, "Taming the Sacramento."

59. U.S. Army Corps of Engineers, South Pacific Division, *Water Resources Development, California, 1979.*

60. Kelley, *Battling the Inland Sea*; Thompson, *The Settlement Geography of the Sacramento-San Joaquin Delta, California.*

61. Kelley, *Battling the Inland Sea*; River Improvement and Drainage Association of California, *Bulletin*, no. 7, 1908, 5–6.

62. "Give This Matter Your Moral Support."

63. Kelley, *Battling the Inland Sea.*

64. LeRoy Percy, son of planter W. A. Percy, became the lawyer of Mississippi state's second levee board for the district in the north Yazoo-Mississippi Delta. He was also lawyer to the Illinois Central Railroad, which had lines in the Delta, and he sat on the board of J. P. Morgan's Southern Railroad. LeRoy Percy became a director of the Federal Reserve Bank and of philanthropic foundations in New York and served as senator from Mississippi. Percy met the president in 1902 on the Delta bear hunt organized by John M. Parker that inspired creation of "Teddy's bear" toys. The hunting party included two cabinet secretaries and Stuyvesant Fish, Illinois Central president, as well as President Roosevelt. Barry, *Rising Tide.*

65. Harrison, *Levee Districts and Levee Building in Mississippi.*

66. Percy helped to organize a large plantation for Cannon co-owned by Senate Appropriations Committee chair William Allison and by the president of the Illinois Central Railroad, Stuyvesant Fish. Percy and Parker were also friends with House Democratic leader John Sharp Williams, a Yazoo-Mississippi Delta planter. Barry, *Rising Tide*; Daniel, *The Shadow of Slavery.*

67. Schapsmeier and Schapsmeier, *Political Parties and Civic Action Groups.* His vice president addressed a meeting in 1908. National Rivers and Harbors Congress, *Proceedings of the Convention*, 1908, 14.

68. Barry, *Rising Tide.*

69. National Rivers and Harbors Congress, *Proceedings of the Convention*, 1908, 29, 198; 1914, 24.

70. Robinson, *Water for the West*; Swain, *Federal Conservation Policy, 1921–1933.*

71. National Rivers and Harbors Congress, *Proceedings of the Convention*, 1907, 29–32; 1912, 251.

72. Swain, *Federal Conservation Policy, 1921–1933.*

73. Arnold, *The Evolution of the 1936 Flood Control Act*; Hays, *Conservation and the Gospel of Efficiency*; Petulla, *American Environmental History*; Pisani, *From the Family Farm to Agribusiness*; Robbins, *Our Landed Heritage*; Robinson, *Water for the West*; Swain, *Federal Conservation Policy, 1921–1933.*

74. Robinson, *Water for the West.*

75. U.S. Army Corps of Engineers Lower Mississippi Valley Division, *Water Resources Development in Louisiana, 1989*, xiv; National Rivers and Harbors Congress, *Proceedings of the Convention*, 1909, 87.

76. Morgan, *Dams and Other Disasters*; Petulla, *American Environmental History*; Robinson, *Water for the West*; U.S. Army Corps of Engineers Historical Division and Public Affairs Office, *The History of the U.S. Army Corps of Engineers.*

77. Frank, *The Development of the Federal Program of Flood Control on the Mississippi River*; Hays, *Conservation and the Gospel of Efficiency.*

78. National Rivers and Harbors Congress, *Proceedings of the Convention, 1911*, 144.

79. Many involved in the 1902 Sacramento convention forming the Reclamation and Drainage Association (the original River Improvement and Drainage Association) also helped form the renewed River Improvement and Drainage Association in San Francisco in 1904. P. J. Van Loben Sels of Oakland was involved in reclamation south of Sacramento, served on the executive committee of the Reclamation and Drainage Association and on the executive committee of the renewed River Improvement and Drainage Association. Reclamation and Drainage Association, [report]; River Improvement and Drainage Association, *Bulletin*, no. 1, 1904.

Others who helped organize both meetings included Fred W. Zeile of San Francisco, treasurer of the River Improvement and Drainage Association; Fred H. Harvey of Galt; J. W. Snowball of Knights Landing; State Sen. R. T. Devlin, chair of the Republican state committee; George F. McNoble of Stockton, vice president of the 1902 convention and past president of the State Bar Association; and swampland owner John Ferris of San Francisco, the first California delegate to attend a NRHC convention in Washington, D.C., in 1906. Kelley, *Battling the Inland Sea*; Reclamation and Drainage Association, [report]; River Improvement and Drainage Association, *Bulletin*, no. 1, 1904.

Rio Vista banker Peter Cook was an official of large reclamation companies, a member of the general committee of the Reclamation and Drainage Association, vice president of the San Joaquin and Sacramento River Improvement Association, on the general committee of the renewed River Improvement and Drainage Association, and a member of the Sacramento Drainage Commission. Reclamation and Drainage Association, [report]; River Improvement and Drainage Association, *Bulletin*, no. 1, 1904; River Improvement and Drainage Association, *Bulletin*, no. 7, 1908; Kelley, *Battling the Inland Sea*; San Joaquin and Sacramento River Improvement Association, *Articles of Association and By-laws.*

Jesse Poundstone, a major swampland owner who helped organize the 1902 Sacramento convention, was president of Reclamation District 108 and was Colusa County representative on the Sacramento Drainage Commission. Kelley, *Battling the Inland Sea*; River Improvement and Drainage Association, *Bulletin*, no. 7, 1908. A. T. J. Reynolds of Walnut Grove served on the general committee of the Sacramento meeting and on the executive committee of the San Joaquin and Sacramento River Improvement Association. Reclamation and Drainage Association, [report]; San Joaquin and Sacramento River Improvement Association, *Articles of Association and By-laws.*

80. Jennings, "Shall Man or Rivers Rule?" 182, 184.

81. Editorial, *Stockton Record,* 10 February 1911; San Joaquin and Sacramento River Improvement Association, *Articles of Association and By-laws.*

82. Hagwood, *From North Bloomfield to North Fork;* U.S. Army Corps of Engineers, Chief of Engineers, Annual Report, 1918.

83. Editorial, *Nevada City Daily Transcript,* 14 February 1911; Editorial, *Grass Valley Tidings,* 24 February 1911.

84. Kelley, *Battling the Inland Sea.*

85. Kahrl, *The California Water Atlas.*

86. Kelley, *Battling the Inland Sea;* U.S. Army Corps of Engineers, South Pacific Division, *Water Resources Development, California, 1979.*

87. 25 Stat. Cal. Extra Sess., 1911, 1170; Kelley, *Battling the Inland Sea;* Kelley, "Taming the Sacramento."

88. California State Engineer, *Report of the State Engineer of the State of California.*

89. 170 Stat. Cal, 1913, 2520; Barton, *Memorandum on Reclamation Districts Within the Sacramento and San Joaquin Drainage District;* Kelley, *Battling the Inland Sea.*

90. Kelley, *Battling the Inland Sea;* Thompson, *The Settlement Geography of the Sacramento-San Joaquin Delta, California.*

91. National Rivers and Harbors Congress, *Proceedings of the Convention,* 1908, 198; 1909, 10–14; 1910, 12; 1912, 10.

92. National Rivers and Harbors Congress, *Proceedings of the Convention,* 1914, 23; National Drainage Congress, *Official Proceedings of the Sixth Annual Meeting of the National Drainage Congress, Cairo, Illinois, January 19, 20 and 21, 1916* (hereafter cited as National Drainage Congress, *Proceedings, 1916*).

93. National Rivers and Harbors Congress, *Proceedings of the Convention,* 1909, 107.

94. National Rivers and Harbors Congress, *Proceedings of the Convention,* 1908, 112; 1909, 87; 1911, 144; 1914, 185; 1911; 1922a; 1922b; 1924, 87, 252; 1926, 9, 72; 1927, 9, 86; 1928, 11.

95. Reichley, *The Life of the Parties.*

96. National Rivers and Harbors Congress, *Proceedings of the Convention,* 1906; 1907, 137; 1908, 65, 83; 1910, 17; 1911, 13; 1913, 108.

97. National Rivers and Harbors Congress, *Proceedings of the Convention,* 1910, 24; 1911, 7.

98. U.S. Army Corps of Engineers Historical Division and Public Affairs Office, *The History of the U.S. Army Corps of Engineers.*

99. Camillo and Pearcy, *Upon Their Shoulders;* Harrison, *Levee Districts and Levee Building in Mississippi.*

100. National Rivers and Harbors Congress, *Proceedings of the Convention*, 1911; 1912, 143–49.

101. National Rivers and Harbors Congress, *Proceedings of the Convention*, 1912, 221–27.

102. Harrison, *Levee Districts and Levee Building in Mississippi*.

103. Louisiana Engineering Society, *Government Control With Cooperation of Riparian States and Cities in the Construction and Maintenance of Levees, Bank Revetments and Channel Enlargements for the Rivers of the Mississippi Valley As Adequate Means of Flood Protection Resulting In Increased Wealth and Stability of Investment*.

104. Park, *Conservation, Reclamation, Navigation*.

105. Frank, *The Development of the Federal Program of Flood Control on the Mississippi River*.

106. National Rivers and Harbors Congress, *Proceedings of the Convention*, 1912, 55, 146.

107. Camillo and Pearcy, *Upon Their Shoulders*; Fox, *A National Duty. Mississippi River Flood Problem: How the Floods Can Be Prevented*; National Rivers and Harbors Congress, *Proceedings of the Convention*, 1914, 196–98.

108. Kelley, *Battling the Inland Sea*.

109. Frank, *The Development of the Federal Program of Flood Control on the Mississippi River*.

110. Harrison, *Levee Districts and Levee Building in Mississippi*.

111. Fox, *A National Duty*; Mississippi River Levee Association, *Mississippi River Levee Association*; MRILC, "Resolutions."

112. Frank, *The Development of the Federal Program of Flood Control on the Mississippi River*.

113. Fox, *A National Duty*; Mississippi River Levee Association, *Mississippi River Levee Association*.

114. Caldwell, *Flood Control on the Lower Mississippi River*; Fox, *A National Duty*; Frank, *The Development of the Federal Program of Flood Control on the Mississippi River*.

115. National Rivers and Harbors Congress, *Proceedings of the Convention*, 1914, 195, 199. MRLA representatives also spoke at NRHC conventions in 1908, 1912, and 1914. National Rivers and Harbors Congress, *Proceedings of the Convention*, 1908, 191; 1912, 266; 1914, 194, 354.

116. Frank, *The Development of the Federal Program of Flood Control on the Mississippi River*; Mississippi River Levee Association, *Flood Protection on the Lower Mississippi River, A National Problem*; Mississippi River Levee Association, *Prominent Bankers Favoring Legislation by Congress to Prevent Floods*; Mississippi River Levee Association, *Protection from Overflow and Reclamation of Thirty Thousand Square Miles of Alluvial Lands of the Lower Mississippi River, a National Work*.

117. Fox, *A National Duty*, 28–29.

118. Frank, *The Development of the Federal Program of Flood Control on the Mississippi River.*

119. National Drainage Congress, *Proceedings, 1916.*

120. National Drainage Congress, *Official Proceedings of the Eighth Annual Meeting of the National Drainage Congress at St. Louis, Missouri, November 11, 12 and 13, 1919* (hereafter cited as National Drainage Congress, *Proceedings, 1919*); National Drainage Congress, *Proceedings, 1916.*

121. National Drainage Congress, *Proceedings, 1916.*

122. Ibid.

123. Pisani, *From the Family Farm to Agribusiness*; Walton, *Western Times and Water Wars*, 194, on Owens Valley in southern California.

124. Stokes, "The Congressional Pork Barrel."

125. Frank, *The Development of the Federal Program of Flood Control on the Mississippi River*; National Drainage Congress, *Proceedings, 1916.*

126. Quoted in National Rivers and Harbors Congress, *Proceedings of the Convention*, 1914, 20, 31–32, 226–27.

127. Interlocks created by Yazoo-Mississippi Delta planters Charles L. Scott and LeRoy Percy are discussed above. Executives of the Illinois Central Railroad, which had been born of government largesse, as a recipient of federal and state land grants, were closely involved with Scott and Percy and spoke at the Lakes-to-the-Gulf Deep Waterway Association, attended the National Drainage Congress convention, and served on the board of directors of the MRLA. Armstrong, Robinson, and Hoy, *History of Public Works in the United States, 1776–1976*; Barry, *Rising Tide*; Fox, *A National Duty*; National Drainage Congress, *Proceedings, 1916*; Park, *Conservation, Reclamation, Navigation.*

Fox created the most important links among the working staffs of lobbying organizations during this period. He was secretary-manager of the Mississippi River Levee Association (the organization most influential in gaining the 1917 Flood Control Act) and a director in the National Drainage Congress in 1919 (which worked for the Drainage Act and for flood control work by the Reclamation Service), and he gave speeches and raised money for the National Rivers and Harbors Congress in 1909, 1911, and 1913 as special director (the organization most influential in gaining the 1928 and 1936 Flood Control Acts). Fox, "Economic Value of Improved Waterways"; Fox, *A National Duty*; National Drainage Congress, *Proceedings, 1919*; National Rivers and Harbors Congress, *Proceedings of the Convention*, 1908, 191.

George Parsons, the mayor of Cairo, Illinois, located near the confluence of the Ohio and the Mississippi rivers, was on the board of directors of the MRLA and became the president of the National Drainage Congress, serving from 1916 to 1917. Fox, *A National Duty*, 7; Mississippi River Levee Association, *Mississippi River Levee Association*; National Drainage Congress, *Proceedings, 1916*, 3. Frank B. Knight was NDC president in 1916. As chair of the executive committee in 1919, Knight worked with other NDC leaders, with the Chicago

Association of Commerce, and with business people in New Orleans to found a pro-growth commercial association that would also promote flood control aid, the Mississippi Valley Association. Knight became chair of the MVA's Flood Prevention Committee. National Drainage Congress, *Proceedings, 1919*, 33.

Californians in the NRHC included prominent businessman P. J. van Loben Sels, former California Gov. George C. Pardee, Sen. Frank Flint, and Rep. John E. Raker. National Rivers and Harbors Congress, *Proceedings of the Convention*, 1906, 34; 1907, 59; 1910, 236; 1914, 226. C. E. Grunsky worked on irrigation and flood control systems for California land developers and for the California state government, co-wrote a flood control plan that eventually became the blueprint for the Sacramento valley, and served as California vice president to the NRHC in 1908 and as California committeeman on the executive committee of the NDC from 1914 to 1917. Kelley, *Battling the Inland Sea*; National Drainage Congress, *Proceedings, 1916*, 4; National Rivers and Harbors Congress, *Proceedings of the Convention*, 1908, 6.

The president of the Lakes-to-the-Gulf Deep Waterways Association, W. K. Kavanaugh, served as a member of the board of the NRHC for many years. National Rivers and Harbors Congress, *Proceedings of the Convention*, 1910, 28. J. Hampton Moore, longtime president of the Atlantic Deeper Waterways Association and a House representative, was also a member of the NRHC, and he chaired an NRHC session at the 1927 meeting. National Rivers and Harbors Congress, *Proceedings of the Convention*, 1912, 221; 1914, 59; 1925, 59; 1927, 107.

W. G. Waldo was consulting engineer to the Tennessee River Improvement Association and member of the NRHC Committee of Five that was commissioned to write the first set of resolutions on flood control for the NRHC. National Rivers and Harbors Congress, *Proceedings of the Convention*, 1922 (March), 83; 1925, 162–65.

Rep. B. G. Humphreys of Mississippi was a member of the NRHC, and he compiled a book detailing the limitations of the Mississippi River Commission's current mandate and providing data on the value of commerce and land in the lower valley threatened by floods. The MRLA published the book and distributed it widely. He wrote the bill in 1913 that became the basis for the 1917 Flood Control Act, co-sponsored in the Senate by NRHC president Sen. Joseph Ransdell. Humphreys, *Floods and Levees of the Mississippi River*; National Rivers and Harbors Congress, *Bulletin*, no. 1, 1920, 2; Townsend, *Flood Control of the Mississippi River*, 15.

128. Sen. Joseph E. Ransdell of Louisiana was NRHC president from 1907 to 1919; Rep. Riley J. Wilson of Louisiana was NRHC president and ranking Democrat on the House Flood Control Committee from 1926 to 1927, and a powerful New Orleans boss; Mayor Martin Behrman, was on the NRHC Board of Directors for years. Barry, *Rising Tide*; National Rivers and Harbors Congress, *Bulletin*, no. 1, 1920, 2; National Rivers and Harbors Congress, *Proceedings of the*

Convention, 1910, 4; 1926, 85; 1927, 85. The president of the Atlantic Deeper Waterways Association and member of Congress J. Hampton Moore protested to a NRHC convention, "for heaven's sake, men of the Middle West, be reasonable—be reasonable with the people of the East and give us a chance to get something back." National Rivers and Harbors Congress, *Proceedings of the Convention*, 1912, 227.

129. Fox, *A National Duty*.

130. National Rivers and Harbors Congress, *Proceedings of the Convention*, 1909, 160. The National Drainage Association's president, Florida Gov. N. B. Broward, spoke at the 1908 convention. National Rivers and Harbors Congress, *Proceedings of the Convention*, 1908, 165.

131. National Rivers and Harbors Congress, *Proceedings of the Convention*, 1906, 97; 1909, 92; 1910, 28; 1912, 221; 1925, 59.

132. Harrison, *Levee Districts and Levee Building in Mississippi*.

133. New York Board of Trade and Transportation, *Mississippi Valley Floods Relief*.

134. Flood Commission of Pittsburgh, *Report of the Flood Commission of Pittsburgh, Penna*; National Drainage Congress, *Proceedings, 1916*.

135. Townsend, *Flood Control of the Mississippi River*.

136. Mississippi River Levee Association, *Flood Protection*.

137. Camillo and Pearcy, *Upon Their Shoulders*.

138. Frank, *The Development of the Federal Program of Flood Control on the Mississippi River*.

139. Camillo and Pearcy, *Upon Their Shoulders*; Elliott, *The Improvement of the Lower Mississippi River for Flood Control and Navigation*; Fox, *A National Duty*; Kelley, *Battling the Inland Sea*; Kelley, "Taming the Sacramento."

140. Frank, *The Development of the Federal Program of Flood Control on the Mississippi River*.

141. Arnold, *The Evolution of the 1936 Flood Control Act*; Frank, *The Development of the Federal Program of Flood Control on the Mississippi River*.

142. National Drainage Congress, *Proceedings, 1916*.

143. Barry, *Rising Tide*.

144. Harrison, *Levee Districts and Levee Building in Mississippi*; Humphreys, *Floods and Levees of the Mississippi River*; National Rivers and Harbors Congress, *Bulletin*, no. 1, 1920, 2; Townsend, *Flood Control of the Mississippi River*.

145. Camillo and Pearcy, *Upon Their Shoulders*; Frank, *The Development of the Federal Program of Flood Control on the Mississippi River*.

146. Quoted in Frank, *The Development of the Federal Program of Flood Control on the Mississippi River*, 76.

147. Swain, *Federal Conservation Policy, 1921–1933*.

148. Quoted in Frank, *The Development of the Federal Program of Flood Control on the Mississippi River*, 142.

149. Camillo and Pearcy, *Upon Their Shoulders.*

150. Frank, *The Development of the Federal Program of Flood Control on the Mississippi River.*

151. Barry, *Rising Tide.*

152. Harrison, *Levee Districts and Levee Building in Mississippi.*

153. March 1, 1917, c. 144, 39 Stat. 948–51. Camillo and Pearcy, *Upon Their Shoulders*; Frank, *The Development of the Federal Program of Flood Control on the Mississippi River.*

154. Frank, *The Development of the Federal Program of Flood Control on the Mississippi River.*

155. Barry, *Rising Tide.*

156. Armstrong, Robinson, and Hoy, *History of Public Works in the United States, 1776–1976*; Frank, *The Development of the Federal Program of Flood Control on the Mississippi River.*

157. Kelley, *Battling the Inland Sea.*

158. Harrison, *Levee Districts and Levee Building in Mississippi.*

159. Frank, *The Development of the Federal Program of Flood Control on the Mississippi River.*

9 The Fully Designed River

1. Hansen, *Gaining Access*; Reichley, *The Life of the Parties.*

2. August 8, 1917, c. 49, 39 Stat. 250–70; Arnold, *The Evolution of the 1936 Flood Control Act*; Hays, *Conservation and the Gospel of Efficiency*; Swain, *Federal Conservation Policy, 1921–1933.*

3. Dick, *The Lure of the Land*; Hundley, *Water and the West*; National Drainage Congress, *Proceedings, 1916*; Robinson, *Water for the West.*

4. Arnold, *The Evolution of the 1936 Flood Control Act*; Swain, *Federal Conservation Policy, 1921–1933.*

5. Frank, *The Development of the Federal Program of Flood Control on the Mississippi River*; National Rivers and Harbors Congress, *Proceedings of the Convention,* 1922 (March), 1922 (December).

6. Harrison, *Levee Districts and Levee Building in Mississippi*; Kelley, *Battling the Inland Sea*; Thompson, *The Settlement Geography of the Sacramento-San Joaquin Delta, California.*

7. RIDA, *Report.*

8. Kelley, *Battling the Inland Sea*; Thompson and Dutra, *The Tule Breakers.*

9. Sacramento Valley Anti-Debris Association, *A Brief Outline of the Economic Features of Hydraulic Mining.*

10. Kahrl, *The California Water Atlas.*

11. Kelley, *Battling the Inland Sea*; Liebman, *California Farmland*; Thompson, *The Settlement Geography of the Sacramento-San Joaquin Delta, California.*

12. Kahrl, *The California Water Atlas*; Kelley, *Battling the Inland Sea*; Thompson, *The Settlement Geography of the Sacramento-San Joaquin Delta, California*; Thompson and Dutra, *The Tule Breakers.*

13. Frank, *The Development of the Federal Program of Flood Control on the Mississippi River.*

14. Although the data that Humphreys and Abbot presented indicated that outlets lowered the floodplain of the river, the authors recommended against maintaining outlets. Humphreys and Abbot, "Report Upon the Physics and Hydraulics of the Mississippi River." Humphreys may have made this recommendation in order to counter the rival report by Ellet, which called for outlets. Barry, *Rising Tide.* Empirical tests over time contradicted a variety of elements of the Humphreys and Abbot report, particularly the idea that concentrating the flow would scour the channel. The report dominated Corps practice for so long because it was so thorough and because the levees-only policy—which claimed much less land than reservoirs or floodways—was politically more acceptable. Reuss, "Andrew A. Humphreys and the Development of Hydraulic Engineering." Considering that channel conditions were primarily important for navigation purposes, the version of the "levees-only" policy promoted by the commission primarily served the Corps' historic commitment to navigation. Camillo and Pearcy, *Upon Their Shoulders.*

15. Barry, *Rising Tide*; Camillo and Pearcy, *Upon Their Shoulders*; Taylor, "The Subjugation of the Mississippi."

16. Barry, *Rising Tide*; Harrison, *Levee Districts and Levee Building in Mississippi*; Louisiana Engineering Society, *Government Control With Cooperation of Riparian States and Cities in the Construction and Maintenance of Levees, Bank Revetments and Channel Enlargements for the Rivers of the Mississippi Valley as Adequate Means of Flood Protection Resulting in Increased Wealth and Stability of Investment.*

17. Barry, *Rising Tide*; Camillo and Pearcy, *Upon Their Shoulders.*

18. Harrison, *Levee Districts and Levee Building in Mississippi.*

19. Camillo and Pearcy, *Upon Their Shoulders*; New York Board of Trade and Transportation, *Mississippi Valley Floods Relief.*

20. Townsend, *Flood Control of the Mississippi River.*

21. Harrison, *Levee Districts and Levee Building in Mississippi.*

22. Ferguson, *History of the Improvement of the Lower Mississippi River for Flood Control and Navigation, 1932–1939.*

23. Frank, *The Development of the Federal Program of Flood Control on the Mississippi River.*

24. Barry, *Rising Tide.*

25. Ibid.; Camillo and Pearcy, *Upon Their Shoulders.*

26. Barry, *Rising Tide*; Frank, *The Development of the Federal Program of Flood Control on the Mississippi River*; Parker, "Curbing the Mississippi"; Swain, *Federal Conservation Policy, 1921–1933.*

27. National Rivers and Harbors Congress, *Proceedings of the Convention, 1924*, 69–70.

28. Swain, *Federal Conservation Policy, 1921–1933*.

29. Frank, *The Development of the Federal Program of Flood Control on the Mississippi River*.

30. Quoted in Barry, *Rising Tide*, 168.

31. Ibid.

32. Ibid.

33. Frank, *The Development of the Federal Program of Flood Control on the Mississippi River*.

34. National Drainage Congress, *Proceedings, 1916*.

35. Frank, *The Development of the Federal Program of Flood Control on the Mississippi River*, 133.

36. Kahrl, *The California Water Atlas*; Kelley, *Battling the Inland Sea*.

37. National Rivers and Harbors Congress, *Proceedings of the Convention*, 1925, 162–65; 1926, 140–45.

38. National Rivers and Harbors Congress, *Proceedings of the Convention*, 1923, 258; 1924, 248; 1925, 69, 93, 119; 1926, 145; 1927, 24.

39. National Rivers and Harbors Congress, *Proceedings of the Convention*, 1925, 118–23.

40. National Rivers and Harbors Congress, *Proceedings of the Convention*, 1923, 177–78.

41. National Rivers and Harbors Congress, *Proceedings of the Convention*, 1922 (December), 145; 1924, 69–70.

42. National Rivers and Harbors Congress, *Bulletin*, no. 1, 1920, 7–8.

43. National Rivers and Harbors Congress, *Bulletin*, no. 1, 1923.

44. National Rivers and Harbors Congress, *Bulletin*, no. 1, 1926, 11.

45. National Rivers and Harbors Congress, *Proceedings of the Convention*, 1924, 26–27.

46. National Rivers and Harbors Congress, *Proceedings of the Convention*, 1925, 206.

47. U.S. Army Corps of Engineers Historical Division and Public Affairs Office, *The History of the U.S. Army Corps of Engineers*.

48. U.S. Engineer Department, *Estimates of Cost of Examinations, etc. of Streams Where Power Development Appears Feasible*.

49. Swain, *Federal Conservation Policy, 1921–1933*.

50. Camillo and Pearcy, *Upon Their Shoulders*; Harrison, *Levee Districts and Levee Building in Mississippi*; Reuss, *Designing the Bayous*.

51. Barry, *Rising Tide*; Elliott, *The Improvement of the Lower Mississippi River for Flood Control and Navigation*; Swain, *Federal Conservation Policy, 1921–1933*.

52. National Rivers and Harbors Congress, *Proceedings of the Convention*, 1926, 72, 185, 200.

53. Barry, *Rising Tide*; Frank, *The Development of the Federal Program of Flood Control on the Mississippi River.*

54. Barry, *Rising Tide.*

55. Ibid.; McPhee, "Atchafalaya."

56. Barry, *Rising Tide.*

57. Swain, *Federal Conservation Policy, 1921–1933.*

58. Ayres, "What's Left From the Great Flood of '93"; Nasar, "Damage Estimates of Midwest Flood Climbing Sharply."

59. Elliott, *The Improvement of the Lower Mississippi River for Flood Control and Navigation,* plate 36.

60. Barry, *Rising Tide*; Harrison, *Levee Districts and Levee Building in Mississippi.*

61. Frank, *The Development of the Federal Program of Flood Control on the Mississippi River*; Harrison, *Levee Districts and Levee Building in Mississippi*; McPhee, "Atchafalaya."

62. Barry, *Rising Tide*; Frank, *The Development of the Federal Program of Flood Control on the Mississippi River.*

63. Barry, *Rising Tide*; Swain, *Federal Conservation Policy, 1921–1933.*

64. National Rivers and Harbors Congress, *Proceedings of the Convention,* 1925, 22; 1926, 182.

65. Barry, *Rising Tide.*

66. Ibid.; Camillo and Pearcy, *Upon Their Shoulders.*

67. Barry, *Rising Tide*; Daniel, *The Shadow of Slavery.*

68. Barry, *Rising Tide.*

69. Ibid.; National Rivers and Harbors Congress, *Bulletin,* no. 1, 1920, 2; National Rivers and Harbors Congress, *Proceedings of the Convention,* 1927, 85.

70. Ibid.; Frank, *The Development of the Federal Program of Flood Control on the Mississippi River.*

71. Barry, *Rising Tide*; Frank, *The Development of the Federal Program of Flood Control on the Mississippi River.*

72. Barry, *Rising Tide*; Camillo and Pearcy, *Upon Their Shoulders*; Pearcy, "After the Flood: A History of the 1928 Flood Control Act."

73. Frank, *The Development of the Federal Program of Flood Control on the Mississippi River.*

74. Arnold, *The Evolution of the 1936 Flood Control Act.*

75. Barry, *Rising Tide*; Frank, *The Development of the Federal Program of Flood Control on the Mississippi River.*

76. Frank, *The Development of the Federal Program of Flood Control on the Mississippi River.*

77. Camillo and Pearcy, *Upon Their Shoulders*; Pearcy, "After the Flood: A History of the 1928 Flood Control Act." Camillo and Pearcy argue that the

evidence does not support the conclusions of Barry and other historians who portray southern politicians as most responsible for the 1928 plan. Barry, *Rising Tide.*

78. Harrison, *Levee Districts and Levee Building in Mississippi.*

79. Quoted in Swain, *Federal Conservation Policy, 1921–1933*, 108–9; Barry, *Rising Tide*; Camillo and Pearcy, *Upon Their Shoulders*; Pearcy, "After the Flood: A History of the 1928 Flood Control Act."

80. Barry, *Rising Tide.*

81. Camillo and Pearcy, *Upon Their Shoulders*; Elliott, *The Improvement of the Lower Mississippi River for Flood Control and Navigation*, 297, plate 70; National Rivers and Harbors Congress, *Proceedings of the Convention* 1927, 108–22; 1928, 112; Swain, *Federal Conservation Policy, 1921–1933.*

82. National Rivers and Harbors Congress, *Bulletin*, no. 1, 1927; National Rivers and Harbors Congress, *Proceedings of the Convention*, 1927, 35–36, 65–67, 93–105.

83. Swain, *Federal Conservation Policy, 1921–1933.*

84. National Rivers and Harbors Congress, *Bulletin*, no. 1, 1928, 15; National Rivers and Harbors Congress, *Proceedings of the Convention*, 1928, 146–48.

85. Barry, *Rising Tide.*

86. Camillo and Pearcy, *Upon Their Shoulders*; Pearcy, "After the Flood"; Swain, *Federal Conservation Policy, 1921–1933.*

87. Arnold, *The Evolution of the 1936 Flood Control Act.*

88. Barry, *Rising Tide.*

89. Arnold, *The Evolution of the 1936 Flood Control Act.*

90. Camillo and Pearcy, *Upon Their Shoulders*; Frank, *The Development of the Federal Program of Flood Control on the Mississippi River.*

91. Barry, *Rising Tide*; Camillo and Pearcy, *Upon Their Shoulders*; Reuss, *Designing the Bayous.*

92. Barry, *Rising Tide*; Camillo and Pearcy, *Upon Their Shoulders*; Frank, *The Development of the Federal Program of Flood Control on the Mississippi River.*

93. Barry, *Rising Tide*, 406; Camillo and Pearcy, *Upon Their Shoulders.*

94. National Rivers and Harbors Congress, *Proceedings of the Convention*, 1928, 97–104.

95. May 15, 1928, C. 569, 45 Stat. 534–39; Arnold, *The Evolution of the 1936 Flood Control Act*; Barry, *Rising Tide.*

96. "Policy to fit a sectional crisis becomes national, however, and in 1929, Congress refunded California's contributions of $4,300,000 to the Sacramento project." Hart, "Crisis, Community, and Consent in Water Politics," 518.

97. Barry, *Rising Tide.*

98. Ibid.; Morison, *The Oxford History of the American People.*

99. Aside from enthusiastic endorsements of a Central American canal at the turn of the century, the major parties had not featured public works spend-

ing in their platforms since the Civil War. Congressional Quarterly, *National Party Conventions, 1831–1992.*

100. Fox, *John Muir and His Legacy.*

101. Quoted in Morison, *The Oxford History of the American People*, vol. 3, 292; Arnold, *The Evolution of the 1936 Flood Control Act*; Hundley, *Water and the West*; Mitchell, *Depression Decade*; National Rivers and Harbors Congress, *Bulletin*, no. 1, 1930, 3; Reichley, *The Life of the Parties*; Swain, *Federal Conservation Policy, 1921–1933.*

102. Hart, "Crisis, Community, and Consent in Water Politics"; Swain, *Federal Conservation Policy, 1921–1933.*

103. Arnold, *The Evolution of the 1936 Flood Control Act.*

104. Swain, *Federal Conservation Policy, 1921–1933.*

105. Congressional Quarterly, *National Party Conventions, 1831–1992*; Reichley, *The Life of the Parties.*

10 Nationwide Program for Flood Control

1. Shanks, *This Land Is Your Land.*

2. Finegold and Skocpol, *State and Party in America's New Deal*; Goldfield, "Worker Insurgency, Radical Organization, and New Deal Labor Legislation"; Skocpol, *Protecting Soldiers and Mothers*; Skocpol and Finegold, "State Capacity and Economic Intervention in the Early New Deal"; Skowronek, *Building a New American State.*

3. O'Neill, "Why the TVA Remains Unique."

4. Swain, *Federal Conservation Policy, 1921–1933.* For example, the New Deal program for managing grazing on federal lands copied a practice developed by the Forest Service by involving local beneficiaries in decisions. Petulla, *American Environmental History.* New Deal projects for the comprehensive development of water resources in California and on the Columbia River originated in the 1920s. Robinson, *Water for the West.* New Deal crop price support policies continued policies negotiated in the 1920s between the USDA and the American Farm Bureau Federation. Hansen, *Gaining Access.* The New Deal soil conservation programs were based on the 1911 Weeks Act, which had allowed the government to buy private lands for national forest reserves. Robbins, *Our Landed Heritage.* Some of the most innovative programs promoted by Roosevelt and his advisors, such as the agricultural resettlement programs and new towns programs, were short-lived. Goodwyn, *The Populist Moment.* The TVA was one innovative program that lasted.

5. Burns, *Roosevelt*; Congressional Quarterly, *National Party Conventions, 1831–1992*; Goldman, *Rendezvous With Destiny*; Leuchtenburg, *Franklin D. Roosevelt and the New Deal, 1932–1940*; Reichley, *The Life of the Parties.*

6. Arnold, *The Evolution of the 1936 Flood Control Act*; Petulla, *American Environmental History*.

7. Congressional Quarterly, *National Party Conventions, 1831–1992*.

8. Goldman, *Rendezvous With Destiny*.

9. Burns, *Roosevelt*; Goldman, *Rendezvous With Destiny*; Leuchtenburg, *Franklin D. Roosevelt and the New Deal, 1932–1940*; Mitchell, *Depression Decade*; Petulla, *American Environmental History*.

10. Arnold, *The Evolution of the 1936 Flood Control Act*; Clawson, *New Deal Planning*; Leuchtenburg, *Franklin D. Roosevelt and the New Deal, 1932–1940*; Mitchell, *Depression Decade*; Petulla, *American Environmental History*; Robinson, *Water for the West*.

11. Clapp, *The* TVA; Schaffer, "Managing the Tennessee River."

12. Hubbard, *Origins of the* TVA; Swain, *Federal Conservation Policy, 1921–1933*.

13. National Rivers and Harbors Congress, *Proceedings of the Convention*, 1910, 1922 (March), 1925.

14. Davidson, *The Tennessee*; National Rivers and Harbors Congress, *Proceedings of the Convention*, 1922 (March).

15. Hubbard, *Origins of the* TVA; Swain, *Federal Conservation Policy, 1921–1933*.

16. Clapp, *The* TVA; Hubbard, *Origins of the* TVA; Swain, *Federal Conservation Policy, 1921–1933*.

17. Clapp, *The* TVA; Hubbard, *Origins of the* TVA.

18. Ibid.

19. Arnold, *The Evolution of the 1936 Flood Control Act*; Hubbard, *Origins of the* TVA.

20. Colignon, *Power Plays*; Petulla, *American Environmental History*; Reichley, *The Life of the Parties*.

21. Hubbard, *Origins of the* TVA.

22. Arnold, *The Evolution of the 1936 Flood Control Act*; Colignon, *Power Plays*; Petulla, *American Environmental History*; Reichley, *The Life of the Parties*.

23. Robinson, *Water for the West*.

24. Leuchtenburg, "Roosevelt, Norris, and the 'Seven Little TVA's.' "

25. Hargrove, *Prisoners of Myth*.

26. Shallat, *Structures in the Stream*.

27. Arnold, *The Evolution of the 1936 Flood Control Act*.

28. Ibid.; National Rivers and Harbors Congress, *Bulletin*, no. 1, 1920; National Rivers and Harbors Congress, *Proceedings of the Convention*, 1927.

29. Ibid.

30. Ibid.

31. Ibid.; Christie, "The Mississippi Valley Committee."

32. Arnold, *The Evolution of the 1936 Flood Control Act*.

33. The National Resources Board replaced the temporary National Planning Board, which had coordinated PWA projects. In 1935, it became the National

Resources Committee, and in 1939 it was reorganized as the National Resources Planning Board. Ibid.; Christie, "The Mississippi Valley Committee"; Clawson, *New Deal Planning*; Robinson, *Water for the West.*

34. Harold Ickes had named Morris L. Cooke as head of the Mississippi Valley Committee in 1933 to administer PWA water projects in that region. Christie, "The Mississippi Valley Committee"; Clawson, *New Deal Planning.* In 1935, its staff was merged into the National Resource Committee's Water Planning Committee. Arnold, *The Evolution of the 1936 Flood Control Act*; Clawson, *New Deal Planning.* It became the Water Resources Committee (WRC) and was charged to study rivers in the entire country. Christie, "The Mississippi Valley Committee."

35. Christie, "The Mississippi Valley Committee"; Engelbert, "Federalism and Water Resources Development."

36. Arnold, *The Evolution of the 1936 Flood Control Act*; Clawson, *New Deal Planning.*

37. Burns, *Roosevelt*; Leuchtenburg, *Franklin D. Roosevelt and the New Deal, 1932–1940.*

38. National Rivers and Harbors Congress, *Resolutions Adopted by the Annual Convention of the National Rivers and Harbors Congress*, 1935.

39. Clawson, *New Deal Planning.*

40. Leuchtenburg, *Franklin D. Roosevelt and the New Deal, 1932–1940*; Mitchell, *Depression Decade*; Petulla, *American Environmental History.*

41. Leuchtenburg, *Franklin D. Roosevelt and the New Deal, 1932–1940.*

42. Robinson, *Water for the West.*

43. Mitchell, *Depression Decade.*

44. Arnold, *The Evolution of the 1936 Flood Control Act*; Mitchell, *Depression Decade.*

45. Arnold, *The Evolution of the 1936 Flood Control Act.*

46. Moore, *Injustice*, 377.

47. Mitchell, *Depression Decade.*

48. Arnold, *The Evolution of the 1936 Flood Control Act*; Clawson, *New Deal Planning.*

49. U.S. Army Corps of Engineers, South Pacific Division, *Water Resources Development, California, 1979.*

50. Ferguson, *History of the Improvement of the Lower Mississippi River for Flood Control and Navigation, 1932–1939.*

51. Barry, *Rising Tide*; Camillo and Pearcy, *Upon Their Shoulders*, ch. 11; Elliott, *The Improvement of the Lower Mississippi River for Flood Control and Navigation*; Reuss, *Designing the Bayous*; U.S. Army, Office of the Chief of Engineers, *Control of Floods in the Alluvial Valley of the Lower Mississippi River.*

52. June 15, 1936, 49 Stat., c. 548, 1508–13; Ferguson, *History of the Improvement of the Lower Mississippi River.*

53. Camillo and Pearcy, *Upon Their Shoulders.*

54. National Rivers and Harbors Congress, *Resolutions Adopted by the Annual Convention of the National Rivers and Harbors Congress,* 1935.

55. Burns, *Roosevelt*; Christie, "The Mississippi Valley Committee"; Leuchtenburg, *Franklin D. Roosevelt and the New Deal, 1932–1940.*

56. Arnold, *The Evolution of the 1936 Flood Control Act.*

57. Ibid.

58. Ibid.

59. Ibid.

60. Ibid.

61. Ibid.; National Rivers and Harbors Congress, *Resolutions Adopted by the Annual Convention of the National Rivers and Harbors Congress,* 1936.

62. National Rivers and Harbors Congress, *Resolutions Adopted by the Annual Convention of the National Rivers and Harbors Congress,* 1936.

63. Maass, *Muddy Waters*; National Rivers and Harbors Congress, *Resolutions Adopted by the Annual Convention of the National Rivers and Harbors Congress,* 1935.

64. Ibid.

65. Burns, *Roosevelt.*

66. Finegold and Skocpol, *State and Party in America's New Deal.*

67. Young, "The Origins of New Deal Agricultural Policy."

68. Arnold, *The Evolution of the 1936 Flood Control Act*; Arnold, "The Flood Control Act of 1936." Arnold's essay describes Cooke's interest in promoting federal hydropower development and Copeland's efforts to protect private power development as they fought over the bill's amendment.

69. Arnold, *The Evolution of the 1936 Flood Control Act,* viii.

70. Congressional Quarterly, *National Party Conventions, 1831–1992.*

71. June 22, 1936, c. 688, 49 Stat. 1570.

72. Arnold, *The Evolution of the 1936 Flood Control Act.*

73. Baumhoff, *The Dammed Missouri Valley.* Roosevelt suggested in a 1935 report to Congress that he intended to make the National Resources Board a permanent agency, but not while the employment emergency existed. Christie, "The Mississippi Valley Committee." Senator Copeland made a half-hearted effort to amend the 1936 flood control bill and make the National Resources Committee permanent. Months after the nationwide flood control bill passed, the NRC finally submitted its plan for comprehensive development to the president. Arnold, *The Evolution of the 1936 Flood Control Act.* The National Resources Planning Board eventually prepared studies of forty-five major river basins. These informed later work by the Bureau of Reclamation and the Corps of Engineers. Congress ended authorization for the NRPB in 1943. Arnold, *The Evolution of the 1936 Flood Control Act*; Clawson, *New Deal Planning*; Robinson, *Water for the West.*

74. Marion Dickens, president of the White and Black Rivers Flood Control Association in Newport, Arkansas, did write to Roosevelt in early 1936 asking him to back a valley authority for Arkansas. Leuchtenburg, "Roosevelt, Norris, and the 'Seven Little TVA's.'" The Bureau of Reclamation adopted power development primarily to subsidize irrigation work, not to foster integrated, comprehensive resource development.

75. Colignon, *Power Plays.*

76. Arnold, *The Evolution of the 1936 Flood Control Act*; Leuchtenburg, "Roosevelt, Norris, and the 'Seven Little TVA's.'"

77. Leuchtenburg, "Roosevelt, Norris, and the 'Seven Little TVA's.'"

78. Christie, "The Mississippi Valley Committee"; Leuchtenburg, "Roosevelt, Norris, and the 'Seven Little TVA's,'" 419.

79. Leuchtenburg, "Roosevelt, Norris, and the 'Seven Little TVA's.'"

80. Goodwin, "The Valley Authority Idea."

81. Baumhoff, *The Dammed Missouri Valley*; Goodwin, "The Valley Authority Idea"; Maass, *Muddy Waters.*

82. Baumhoff, *The Dammed Missouri Valley*; Goodwin, "The Valley Authority Idea"; Maass, *Muddy Waters.*

83. Leuchtenburg, "Roosevelt, Norris, and the 'Seven Little TVA's'"; Maass, *Muddy Waters.*

84. Arnold, *The Evolution of the 1936 Flood Control Act.*

85. Selznick, TVA *and the Grass Roots.*

86. Arnold, *The Evolution of the 1936 Flood Control Act*; Leuchtenburg, "Roosevelt, Norris, and the 'Seven Little TVA's.'"

87. Christie, "The Mississippi Valley Committee"; Leuchtenburg, "Roosevelt, Norris, and the 'Seven Little TVA's'"; Selznick, TVA *and the Grass Roots.*

88. Leuchtenburg, *Franklin D. Roosevelt and the New Deal, 1932–1940.*

89. Arnold, *The Evolution of the 1936 Flood Control Act*; Hart, "Crisis, Community, and Consent in Water Politics"; Leuchtenburg, "Roosevelt, Norris, and the 'Seven Little TVA's.'"

90. The Boeuf Floodway was formally cancelled in 1941. Camillo and Pearcy, *Upon Their Shoulders*; Ferguson, *History of the Improvement of the Lower Mississippi River*; Reuss, *Designing the Bayous.*

91. Elliott, *The Improvement of the Lower Mississippi River for Flood Control and Navigation.*

92. Ferguson, *History of the Improvement of the Lower Mississippi River*; U.S. Army Corps of Engineers Lower Mississippi Valley Division, *Water Resources Development in Louisiana, 1989.*

93. Harrison, *Levee Districts and Levee Building in Mississippi.*

94. U.S. Army Corps of Engineers Lower Mississippi Valley Division, *Water Resources Development in Louisiana, 1989.*

95. Camillo and Pearcy, *Upon Their Shoulders*; Reuss, *Designing the Bayous.*

96. Arnold, *The Evolution of the 1936 Flood Control Act*; Moore and Moore, *The Army Corps of Engineers and the Evolution of Federal Flood Plain Management Policy.*

97. Worster, *Rivers of Empire.*

98. Ibid.

99. The Corps would provide the water and reservoir facilities at no cost for recipients in the seventeen western states covered by the 1902 Reclamation Act. In other states, state and local governments would assume half of the cost of allocating reservoir capacity to irrigation. U.S. Army Corps of Engineers Lower Mississippi Valley Division, *Water Resources Development in Louisiana, 1989.*

100. Maass, *Muddy Waters*; Reisner, *Cadillac Desert*; Worster, *Rivers of Empire.*

101. Baumhoff, *The Dammed Missouri Valley*; Reisner, *Cadillac Desert*; Robinson, *Water for the West*; Worster, *Rivers of Empire.*

102. Maass, *Muddy Waters*; Morgan, *Dams and Other Disasters.*

103. U.S. Army Corps of Engineers Lower Mississippi Valley Division, *Water Resources Development in Louisiana, 1989.*

104. Arnold, *The Evolution of the 1936 Flood Control Act.*

105. Lowi, "Four Systems of Policy, Politics, and Choice"; Maass, *Muddy Waters.*

106. U.S. Army Corps of Engineers Lower Mississippi Valley Division, *Water Resources Development in Louisiana, 1989.*

107. Arnold, *The Evolution of the 1936 Flood Control Act*; Moore and Moore, *The Army Corps of Engineers and the Evolution of Federal Flood Plain Management Policy.*

108. U.S. Army Corps of Engineers Lower Mississippi Valley Division, *Water Resources Development in Louisiana, 1989.*

109. By mid-century, the Corps was the largest constructor and operator of federal power facilities. U.S. Army Corps of Engineers Historical Division and Public Affairs Office, *The History of the U.S. Army Corps of Engineers.*

110. Gottlieb and FitzSimmons, *Thirst for Growth*; McPhee, "Atchafalaya"; Robinson, *Water for the West.*

111. U.S. Army Corps of Engineers Lower Mississippi Valley Division, *Water Resources Development in Louisiana, 1989.*

112. McPhee, "Atchafalaya"; Stine, "United States Army Corps of Engineers (CE)."

113. Reuss, "The Development of American Water Resources."

114. Arnold, *The Evolution of the 1936 Flood Control Act.*

115. Grunwald, "Working to Please Hill Commanders."

116. Grunwald, "Agency Says Engineers Likely Broke Rules"; Grunwald, "Shooting Pork in the Barrel."

117. Reuss, "The Development of American Water Resources."

118. Buel, "A Race With Nature"; Hoyt and Langbein, *Floods*; Rogers, *Amer-*

ica's Water; White, "Human Adjustment to Floods"; White and Edinger, *Choice of Adjustment to Floods.*

119. Grunwald, "Engineers of Power"; Grunwald, "Working to Please Hill Commanders."

120. U.S. Army Corps of Engineers Lower Mississippi Valley Division, *Water Resources Development in Louisiana, 1989.*

121. Reuss, "The Development of American Water Resources."

122. Ayres, "What's Left From the Great Flood of '93"; Perkins, "Flood Insurance."

123. Arnold, *The Evolution of the 1936 Flood Control Act.*

124. U.S. Army Corps of Engineers Lower Mississippi Valley Division, *Water Resources Development in Louisiana, 1989*, viii.

125. Grunwald, "Working to Please Hill Commanders."

126. Revkin, "Stockpiling Water for a River of Grass."

127. Arnold, *The Evolution of the 1936 Flood Control Act.*

128. Bakker, *An Island Called California.*

129. Rohe, "Man and the Land."

130. Kelley, *Battling the Inland Sea*; U.S. Army Corps of Engineers, South Pacific Division, *Water Resources Development, California, 1979.*

131. Perkins, "Flood Insurance"; Wyatt, "This Year's Flooding Is 1 in 100."

132. Kelley, *Battling the Inland Sea*, 310.

133. Ibid.; Smith, "Deal OK'd for First River Plan Since 1911."

134. Barnum, "Water Warriors Prepare for New Battle in Delta"; McLeod, " 'Historic' Water Pact Rooted in Compromise."

135. Childs, *Mighty Mississippi.*

136. U.S. Army Corps of Engineers Lower Mississippi Valley Division, *Water Resources Development in Louisiana, 1989.*

137. Barry, *Rising Tide*; U.S. Army Corps of Engineers Lower Mississippi Valley Division, *Water Resources Development in Louisiana, 1989.*

138. Childs, *Mighty Mississippi*; Elliott, *The Improvement of the Lower Mississippi River for Flood Control and Navigation*; Harrison, *Levee Districts and Levee Building in Mississippi*; McPhee, "Atchafalaya"; Nasar, "Damage Estimates of Midwest Flood Climbing Sharply"; U.S. Army Corps of Engineers Lower Mississippi Valley Division, *Water Resources Development in Louisiana, 1989.*

139. Ayres, "What's Left From the Great Flood of '93."

140. Camillo and Pearcy, *Upon Their Shoulders*; Ferguson, *History of the Improvement of the Lower Mississippi River.*

141. Barry, *Rising Tide*; McPhee, "Atchafalaya"; Reuss, *Designing the Bayous*; U.S. Army Corps of Engineers Lower Mississippi Valley Division, *Water Resources Development in Louisiana, 1989.*

142. Lt. Gen. Carl Strock, Commander, U.S. Army Corps of Engineers. Defense Department Briefing on Hurricane Katrina Repairs. News Transcript.

Office of the Assistant Secretary of Defense (Public Affairs). September 15, 2005. http://www.defenselink.mil/transcripts/2005/tr20050915-3905.html.

143. Mitchell, "Society, Economy, and the State Effect."

11 Rivers by Design

1. Kelley, "The Context and the Process."

2. Rokkan, *Cities, Elections, Parties.*

3. Sahlins, *Boundaries.*

4. Mitchell, "Society, Economy, and the State Effect," 95.

5. De Roos and Maass, "The Lobby That Can't Be Licked"; Grunwald, "Engineers of Power"; Lowi, "American Business, Public Policy, Case-Studies, and Political Theory"; Lowi, "Four Systems of Policy, Politics, and Choice"; Lowi, "The Welfare State, the New Regulation, and the Rule of Law." Maass, *Muddy Waters*; Morgan, *Dams and Other Disasters.*

6. Stein and Bickers, *Perpetuating the Pork Barrel.*

7. Ibid.; Welsh, "Beyond Designed Capture."

8. McConnell, *The Decline of Agrarian Democracy*; McConnell, *Private Power and American Democracy.*

Bibliography

Abbot, Edwin Hale. *A Review of the Report Upon the Physics and Hydraulics of the Mississippi River.* Boston: Crosby and Nichols, 1862. Reprinted from *North American Review.*

Aiken, Charles S. "The Decline of Sharecropping in the Lower Mississippi River Valley." *Geoscience and Man* 19 (1978): 151–65.

Armstrong, Ellis L., Michael C. Robinson, and Suellen M. Hoy. *History of Public Works in the United States, 1776–1976.* Chicago: American Public Works Association, 1976.

Arnold, Joseph L. *The Evolution of the 1936 Flood Control Act.* Fort Belvoir, Va.: Corps of Engineers, United States Army, 1988.

——. "The Flood Control Act of 1936: A Study in Politics, Planning, and Ideology." In *The Flood Control Challenge: Past, Present, and Future,* edited by Howard Rosen and Martin Reuss, 13–27. Chicago: Public Works Historical Society, 1988.

Ashton, T. H., and C. H. E. Philpin, eds. *The Brenner Debate: Agrarian Class Structure and Economic Development in Pre-Industrial Europe.* Cambridge: Cambridge University Press, 1985.

Ayres, B. Drummond, Jr. "What's Left From the Great Flood of '93." *New York Times,* 10 August 1993, 1, B8.

Bakker, Elna S. *An Island Called California: An Ecological Introduction to Its Natural Communities.* 2nd ed. Berkeley: University of California, 1984.

Bancroft, Hubert Howe. *History of California, 1860–1890.* San Francisco: The History Co., 1890.

Barnum, Alex. "Water Warriors Prepare for New Battle in Delta." *San Francisco Chronicle,* 16 March 1998, A1, A4.

Barry, John M. *Rising Tide: The Great Mississippi Flood of 1927 and How It Changed America.* New York: Simon and Schuster, 1997.

Bartlett; Ruhl. *Policy and Power: Two Centuries of American Foreign Relations.* New York: Hill and Wang, 1963.

Barton, A. M. *Memorandum on Reclamation Districts Within the Sacramento and San Joaquin Drainage District. Read Before the California Commission on Irrigation and Reclamation District Financing.* Sacramento: California Reclamation Board, 1930.

Baumhoff, Richard G. *The Dammed Missouri Valley: One-Sixth of Our Nation.* New York: Alfred A. Knopf, 1951.

Bean, Edwin F. *Bean's History and Directory of Nevada County*. Nevada City, 1867.

Beck, Warren A., and Ynez D. Haase. *Historical Atlas of California*. Norman: University of Oklahoma Press, 1974.

Bensel, Richard Franklin. *Sectionalism and American Political Development, 1880–1980*. Madison: University of Wisconsin Press, 1984.

——. *Yankee Leviathan: The Origins of Central State Authority in America, 1859–1877*. Cambridge: Cambridge University Press, 1990.

Bishofberger, Thomas E. "Early Flood Control in the Central Valley." *Journal of the West* 14 (1975): 85–94.

Bliss, George. Letter Regarding the Rivers and Harbors Bill, October 2, 1882. New York Public Library, 1882.

Brenner, Neil, Bob Jessop, Martin Jones, and Gordon MacLeod. "Introduction: State Space in Question." In *State/Space: A Reader*, edited by Neil Brenner, Bob Jessop, Martin Jones, and Gordon MacLeod, 1–26. Malden, Mass.: Blackwell Publishing, 2003.

Brewer, William H. *Up and Down California in 1860–1864: The Journal of William H. Brewer*. 1864. Reprint edited by Francis P. Farquahar. Berkeley: University of California, 1949.

Brienes, Marvin. "Sacramento Defies the Rivers, 1858–1878." *California History* 57, no. 1 (1979): 2–19.

Buel, Stephen. "A Race With Nature." *San Jose Mercury News*, 18 October 1997, 1, 24.

Burnham, Walter Dean. *Critical Elections and the Mainsprings of American Politics*. New York: W. W. Norton, 1970.

Burns, James MacGregor. *Roosevelt: The Lion and the Fox*. New York: Harcourt, Brace, 1956.

Burton, Ian. "Some Aspects of Flood Loss Reduction in England and Wales." In *Papers on Flood Problems. Department of Geography Research Paper* No. 70, edited by Gilbert F. White. Chicago: Geography Department, University of Chicago, 1961.

Caldwell, Albert S. *Flood Control on the Lower Mississippi River. An Address Delivered Before the Alumni of the University of the South, Sewanee, Tennessee, by Albert S. Caldwell, President of the Mississippi River Levee Association*. Memphis: Mississippi River Levee Association, 1913.

California Commissioner of Public Works. *Report of the Commissioner of Public Works to the Governor of California, Together with the Report of the Commissioner of Public Works Upon the Rectification of the Sacramento and San Joaquin Rivers and Their Principal Tributaries, and the Reclamation of the Overflowed Lands Adjacent Thereto*. Sacramento: California Commissioner of Public Works, 1905.

California State Controller. *Annual Report of the Controller of the State for the 15th and 16th Fiscal Years, 1864 and 1865*. Sacramento: California State Controller, 1866.

California State Engineer. *Report of the State Engineer of the State of California. November 30, 1908 to November 30, 1910*. Sacramento: California State Engineer, 1911.

California Surveyor General. *Annual Report of the Surveyor General of California for the Year 1860*. Sacramento: Surveyor General, 1861.

——. *Annual Report of the Surveyor General of California for the Year 1862*. Sacramento: Surveyor General, 1863.

——. *Biennial Report of the Surveyor General of the State of California, From August 1, 1873 to August 1, 1875*. Sacramento: Surveyor General, 1876.

——. *Statistical Report of the Surveyor General of California for the Years 1869, 1870, and 1871*. Sacramento: Surveyor General, 1872.

Camillo, Charles A., and Matthew T. Pearcy. *Upon Their Shoulders: A History of the Mississippi River Commission From Its Inception Through the Advent of the Modern Mississippi River and Tributaries Project*. Vicksburg, Miss.: Mississippi River Commission, 2004.

Caughey, John W., and Norris Hundley Jr. *California: History of a Remarkable State*. 4th ed. Englewood Cliffs, N.J.: Prentice-Hall, 1982.

Chandler, Alfred N. *Land Title Origins: A Tale of Force and Fraud*. New York: Robert Schalkenbach Foundation, 1945.

Childs, Marquis W. *Mighty Mississippi: Biography of a River*. New Haven: Ticknor and Fields, 1982.

Christie, Jean. "The Mississippi Valley Committee: Conservation and Planning in the Early New Deal." *Historian* 32 (1970): 449–69.

Cincinnati City Council. *Memorial of the Citizens of Cincinnati, to the Congress of the United States, Relative to the Navigation of the Ohio and Mississippi Rivers*. Cincinnati: Cincinnati City Council, 1844.

Clapp, Gordon R. *The* TVA: An Approach to the Development of a Region. Chicago: University of Chicago Press, 1955.

Clawson, Marion. *New Deal Planning: The National Resources Planning Board*. Baltimore: Johns Hopkins University Press for Resources for the Future, 1981.

Colignon, Richard A. *Power Plays: Critical Events in the Institutionalization of the Tennessee Valley Authority*. Albany: State University of New York Press, 1997.

Collins, Randall. "Some Principles of Long-Term Social Change: The Territorial Power of States." In *Research in Social Movements, Conflicts, and Change*, edited by Louis Kriesberg, 1–34. Greenwich, Conn.: JAI Press, 1978.

Commercial Club of St. Louis. *Hon. W. M. Fishback (Governor of Arkansas), Hon. W. C. Stone (Governor of Missouri), and Mr. H. C. Haarstick (President of the St. Louis and Mississippi Valley Transportation Co). Speeches Delivered Before the Commercial Club of St. Louis, November 18, 1893*. St. Louis: St. Louis Commercial Club, 1893.

Commonwealth Club. *Transactions of the Commonwealth Club of California*. Vol. 4, 1909.

Congressional Convention St. Louis. *Proceedings of the Congressional Convention Held in the City of St. Louis on 13th, 14th and 15th days of May 1873.* St. Louis: Executive Committee, Congressional Convention, St. Louis, 1873.

Congressional Quarterly. *National Party Conventions, 1831–1992.* Washington, D.C.: Congressional Quarterly, 1995.

"Convention Assembles to Devise Ways and Means to Prevent Recurrence of the Destructive Overflows." *San Francisco Chronicle,* 24 May 1904.

Convention for the Improvement of the Mississippi River Washington D.C. *Proceedings of the Convention for the Improvement of the Mississippi River and Its Navigable Tributaries, Together With a Brief Memorial to Congress, Held at Washington, D.C., Feb. 5, 6, 7, 1884.* Washington, D.C.: Convention on the Improvement of the Mississippi River, 1884.

Convention on the Improvement of the Western Waterways Cincinnati. *Proceedings of the Convention on the Improvement of the Western Water-Ways, Held at Cincinnati, September 4th and 5th, 1889.* Cincinnati: Convention on the Improvement of the Western Waterways, Cincinnati, Ohio, 1889.

Convention on the Improvement of the Western Waterways Memphis. *Proceedings of the Convention on the Improvement of the Western Water-ways, Together With a Brief Memorial to Congress. Held at Memphis, Tennessee, October 20 and 21, 1887.* Memphis: Convention on the Improvement of the Western Waterways, Memphis, 1887.

Convention on the Improvement of the Western Waterways New Orleans. *Proceedings of the Convention on the Improvement of the Western Waterways, Together With a Brief Memorial to Congress. Held at the World's Industrial and Cotton Cetennial* [sic] *Exposition, New Orleans, Louisiana, April 7th and 8th, 1885.* New Orleans: Convention on the Improvement of the Western Waterways, New Orleans, 1885.

Coulter, E. Merton. "The Efforts of the Democratic Societies of the West to Open the Navigation of the Mississippi." *Mississippi Valley Historical Review* 11, no. 3 (1924): 376–89.

Dana, Julian. *The Sacramento: River of Gold.* New York: Farrar and Rinehart, 1939.

Daniel, Pete. *The Shadow of Slavery: Peonage in the South, 1901–1969.* Urbana: University of Illinois Press, 1972.

Davidson, Donald. *The Tennessee.* Volume 2: *The New River, Civil War to* TVA. New York: Rinehart, 1948.

Davis, S. Rufus. *The Federal Principle: A Journey Through Time in Quest of Meaning.* Berkeley: University of California Press, 1978.

de Roos, Robert, and Arthur A. Maass. "The Lobby That Can't Be Licked." *Harpers* August (1949): 21.

Delegates from the City of St. Louis. *The Commerce and Navigation of the Valley of the Mississippi and Also That Appertaining to the City of St. Louis: Considered*

With Reference to the Improvement By the General Government, of the Mississippi River and Its Principal Tributaries; being a Report, Prepared By Authority of the Delegates From the City of St. Louis, for Use of the Chicago Convention of July 5, 1847. St. Louis: Delegates From the City of St. Louis [1847?].

Dick, Everett. *The Lure of the Land*. Lincoln: University of Nebraska Press, 1970.

Dobbin, Frank. *Forging Industrial Policy: The United States, Britain, and France in the Railway Age*. Cambridge: Cambridge University Press, 1994.

Downey, John G. *Annual Message of Governor John G. Downey*. Sacramento, 1862.

Drake, Charles D. *The Mission of the Mississippi River. Speech of Charles D. Drake, of St. Louis, Delivered October 23, 1851, Before a Convention Held at Burlington, Iowa, to Devise Ways and Means to Procure the aid of Congress in Removing Obstructions to the Navigation of the Mississippi River at the Des Moines and Rock River Rapids*. Burlington, Iowa: Burlington Convention, 1851.

Eads, James Buchanan. *Physics and Hydraulics of the Mississippi River: Report of the United States Levee Commission*. New Orleans: James Eads, 1876.

Editorial. *Grass Valley Tidings*, 24 February 1911.

Editorial. *Nevada City Daily Transcript*, 14 February 1911.

Editorial. *Stockton Record*, 10 February 1911.

Elliott, D. O. *The Improvement of the Lower Mississippi River for Flood Control and Navigation*. Vicksburg, Miss.: U.S. Waterways Experiment Station, Corps of Engineers, U.S. Army, 1932.

Ellis, William T. *Memories: Seventy-two Years in the Romantic County of Yuba*. Eugene: University of Oregon Press, 1939.

Engelbert, Ernest A. "Federalism and Water Resources Development." *Law and Contemporary Problems* 22, no. 3 (1957): 325–50.

Ferguson, H. B. *History of the Improvement of the Lower Mississippi River for Flood Control and Navigation, 1932–1939*. Vicksburg, Miss.: Mississippi River Commission, 1940.

Ferrell, Robert H., and Richard Natkiel. *Atlas of American History*. New York: Facts on File Publications, 1990.

Finegold, Kenneth, and Theda Skocpol. *State and Party in America's New Deal*. Madison: University of Wisconsin Press, 1995.

Flood Commission of Pittsburgh. *Report of the Flood Commission of Pittsburgh, Penna*: Flood Commission of Pittsburgh, 1912.

Foner, Eric. *Reconstruction: America's Unfinished Revolution, 1863–1877*. Cambridge, Mass.: Harper and Row, 1988.

Forshey, Caleb G. *Memoir on the Physics of the Mississippi River, and Certain Internal Improvements in the State of Louisiana. Taken From the Report of the Joint Committee on Levees, Louisiana Legislature*. New Orleans: Printed at the Office of the "Bee," 1850.

Fox, Gregory H. "New Approaches to International Human Rights: The Sovereign State Revisited." In *State Sovereignty: Change and Persistence in International Relations*, edited by Sohail H. Hashmi, 105–30. University Park: Pennsylvania State University Press, 1997.

Fox, John. *John Muir and His Legacy: The American Conservation Movement.* Boston: Little, Brown, 1981.

Fox, John A. "Economic Value of Improved Waterways." *California Weekly*, 21 May 1909, 411–12.

——. *A National Duty. Mississippi River Flood Problem: How the Floods Can be Prevented.* Memphis: Mississippi River Levee Association, 1915.

Frank, Arthur DeWitt. *The Development of the Federal Program of Flood Control on the Mississippi River.* New York: AMS Press, 1968. First published 1930 by Columbia University Press.

Friedman, Lawrence M. *A History of American Law.* 2nd ed. New York: Simon and Schuster, 1985.

Gates, Paul W. *Agriculture and the Civil War.* New York: Alfred A. Knopf, 1965.

——. "The Homestead Law in an Incongruous Land System." *American Historical Review* 41 (1936): 652–81.

——. "The Nationalizing Influence of the Public Lands: Indiana." In *This Land of Ours: The Acquisition and Disposition of the Public Domain (Indiana American Revolution Bicentennial Symposium)*, 103–26. Indianapolis: Indiana Historical Society, 1978.

Gates, Paul W., and Robert W. Swenson. *History of Public Land Law Development.* Washington, D.C.: U.S. Public Land Law Review Commission, 1968.

Gellner, Ernest. *Nations and Nationalism.* Ithaca: Cornell University Press, 1983.

Gen. Cass' Letter to the Harbor and River Convention. New York: People's Press, 1848.

"Give This Matter Your Moral Support." *The Grizzly Bear*, May 1908, 38.

Goldfield, Michael. "Worker Insurgency, Radical Organization, and New Deal Labor Legislation." *American Political Science Review* 83, no. 4 (1989): 1257–82.

Goldman, Eric F. *Rendezvous With Destiny: A History of Modern American Reform.* New York: Vintage Books, 1956.

Goodrich, Carter. *Government Promotion of American Canals and Railroads, 1800–1890.* New York: Columbia University Press, 1960.

Goodwin, Craufurd D. "The Valley Authority Idea: The Fading of a National Vision." In *TVA: Fifty Years of Grass-roots Bureaucracy*, edited by Erwin C. Hargrove and Paul K. Conkin, 263–86. Urbana: University of Illinois Press, 1983.

Goodwyn, Lawrence. *The Populist Moment: A Short History of the Agrarian Revolt in America.* New York: Oxford University Press, 1978.

Gottlieb, Robert, and Margaret FitzSimmons. *Thirst for Growth: Water Agencies as Hidden Government in California.* Tucson: University of Arizona Press, 1991.

Grantham, Dewey W. *The Life and Death of the Solid South: A Political History*. Lexington: University Press of Kentucky, 1988.

Grunwald, Michael. "Agency Says Engineers Likely Broke Rules." *Washington Post*, 29 February 2000, A04.

——. "Engineers of Power: An Agency of Unchecked Clout." *Washington Post*, 10 September 2000, A1.

——. "Shooting Pork in the Barrel: Corps of Engineers Called Part of Wasteful 'Iron Triangle.'" *Washington Post*, 2 March 2000, A17.

——. "Working to Please Hill Commanders." *Washington Post*, 11 September 2000, A01.

Hagwood, Joseph J., Jr. *The California Debris Commission: A History*. Sacramento: U.S. Army Corps of Engineers, Sacramento District, 1981.

——. *Commitment to Excellence: A History of the Sacramento District U.S. Army Corps of Engineers, 1929–1973*. Sacramento: U.S. Army Engineer District, 1976.

——. *From North Bloomfield to North Fork: Attempts to Comply With the Sawyer Decision*. Sacramento: Sacramento State College, Master's thesis, 1970.

Hall, William Ham. Sacramento Valley River Improvement: Government Policy and Works. Addressed to: Col. William H. Heuer, Corps of Engineers. Board of Engineers for Improvement of the Sacramento and Tributary Rivers, 1905. Bancroft Library, University of California, Berkeley.

Hall, William Mosley. *Chicago River-and-Harbor Convention July 5th, 6th, and 7th, 1847: An Account of Its Origin and Proceedings and Statistics*. Edited by Robert Fergus. Chicago: Fergus Printing Co., 1882.

Hansen, John Mark. *Gaining Access: Congress and the Farm Lobby, 1919–1981*. Chicago: University of Chicago Press, 1991.

Harbor and River Convention [of 1847] Chicago. *Memorial to the Congress of the United States of the Executive Committee of the Convention Held at Chicago, July 5, 1847. With an Abstract of the Convention, on the Improvement of Rivers and Harbors. Presented to Congress, June 1848*. Albany: Harbor and River Convention, Chicago, 1848.

Harbor and River Convention Chicago. *Proceedings of the Harbor and River Convention Held at Chicago, July Fifth, 1847: Together With a Full List of Names of Delegates in Attendance: Letters Read at the Convention and a Detailed Appendix Published by the Convention*. Chicago: Harbor and River Convention, 1847.

Hargrove, Erwin C. *Prisoners of Myth: The Leadership of the Tennessee Valley Authority, 1933–1990*. Princeton: Princeton University Press, 1994.

Harris, Marshall. *Origin of the Land Tenure System in the United States*. Ames: Iowa State College Press, 1953.

Harrison, Robert W. *Alluvial Empire: A Study of State and Local Efforts Toward Land Development in the Alluvial Valley of the Lower Mississippi River*. Little Rock, Ark.: Delta Fund and Economic Research Service, USDA, 1961.

Harrison, Robert W. *Levee Districts and Levee Building in Mississippi: A Study of State and Local Efforts to Control Mississippi River Floods.* Washington, D.C.: Bureau of Agricultural Economics, USDA, Delta Council, the Board of Mississippi Levee Commissioners, Board of Levee Commissioners for the Yazoo-Mississippi Delta, and Mississippi Agricultural Experiment Station, Cooperating with the Bureau of Agricultural Economics, U.S. Department of Agriculture, 1951.

Hart, Henry C. "Crisis, Community, and Consent in Water Politics." *Law and Contemporary Problems* 22, no. 3 (1957): 472–509.

Harvey, David. *Justice, Nature and the Geography of Difference.* Oxford: Blackwell Publishers, 1996.

Haupt, Lewis M. W. "The Mississippi River Problem." Pamphlet reprinted from *Proceedings of the American Philosophical Society* 40, no. 175 (1904): 71–96.

Hays, Samuel P. *Conservation and the Gospel of Efficiency: The Progressive Conservation Movement, 1890–1920.* 1959. Reprint, Pittsburgh: University of Pittsburgh Press, 1999.

Hibbard, Benjamin. *A History of the Public Land Policies.* 1924. Reprint, New York: Macmillan, 1965.

Hilgard, Eugene W. *Reclamation of the Alluvial Basin of the Mississippi River. . . Being a Supplement to the Report of the Commission of Engineers of January 16, 1875.* Washington, D.C.: Commission on the Reclamation of the Alluvial Basin of the Mississippi River, 1878.

Horwitz, Morton J. *The Transformation of American Law: 1780–1860.* Cambridge, Mass.: Harvard University Press, 1977.

Hoyt, William G. , and Walter B. Langbein. *Floods.* Princeton: Princeton University Press, 1955.

Hubbard, Preston J. *Origins of the TVA: The Muscle Shoals Controversy, 1920–1932.* New York: W. W. Norton, 1961.

Humphreys, Andrew A., and Henry L. Abbot. *Report Upon the Physics and Hydraulics of the Mississippi River.* Washington, D.C.: United States War Department, 1861.

Humphreys, Benjamin G. *Floods and Levees of the Mississippi River.* Memphis: Mississippi River Levee Association, 1914.

Hundley, Norris, Jr. *Water and the West: The Colorado River Compact and the Politics of Water in the American West.* Berkeley: University of California Press, 1975.

Jennings, Rufus P. "Shall Man or Rivers Rule?" *California Weekly,* 12 February 1909, 182, 184.

Jewell, Harold W. *The Development of Governmental Restraints on Hydraulic Mining in California.* Sacramento: Sacramento State College, Master's thesis, 1955.

Johnson, Stephen, Gerald Haslam, and Robert Dawson. *The Great Central Val-*

ley: California's Heartland. Berkeley: University of California Press in association with the California Academy of Sciences, 1993.

Kahrl, William L. *The California Water Atlas.* Sacramento: State of California, 1979.

Kane, Anne, and Michael Mann. "A Theory of Early Twentieth-Century Agrarian Politics." *Social Science History* 16, no. 3 (1992): 421–54.

Keller, Morton. *Affairs of State: Public Life in Late Nineteenth Century America.* Cambridge: Belknap Press of Harvard University Press, 1977.

Kelley, Robert. *Battling the Inland Sea: American Political Culture, Public Policy, and the Sacramento Valley, 1850–1986.* Berkeley: University of California Press, 1989.

——. "The Context and the Process: How They Have Changed Over Time." In *Water Resources Administration in the United States,* edited by Martin Reuss, 10–22. East Lansing: American Water Resources Association and Michigan State University Press, 1993.

——. "Forgotten Giant: The Hydraulic Gold Mining Industry in California." *Pacific Historical Review* 23, no. 4 (1954): 343–56.

——. *Gold vs. Grain: The Mining Debris Controversy.* Glendale, Calif.: Arthur H. Clark Co., 1959.

——. "The Mining Debris Controversy in the Sacramento Valley." *Pacific Historical Review* 25, no. 4 (1956): 331–46.

——. "Taming the Sacramento: Hamiltonianism." *Pacific Historical Review* 34 (1965): 21–49.

Kinyon, Edmund. *The Northern Mines.* Grass Valley, Calif: Union Publishing, 1949.

Kulikoff, Allan. *The Agrarian Origins of American Capitalism.* Charlottesville: University Press of Virginia, 1992.

Lakes-to-the-Gulf Deep Waterways Association. *The Deep Waterway Between the Great Lakes and the Gulf of Mexico.* St. Louis: Lakes-to-the-Gulf Deep Waterways Association, 1911.

Laumann, Edward O., and David Knoke. *The Organizational State: Social Choice in National Policy Domains.* Madison: University of Wisconsin Press, 1987.

Lee, Lawrence B. *Reclaiming the American West: An Historiography and Guide.* Santa Barbara: ABC-Clio, 1980.

Lefebvre, Henri. *The Production of Space.* 1974. Translated by Donald Nicholson-Smith. Oxford: Blackwell, 1991.

Leivesley, Sally. "Natural Disasters in Australia." *Disasters* 8, no. 2 (1984): 83–88.

Leuchtenburg, William E. *Franklin D. Roosevelt and the New Deal, 1932–1940.* New York: Harper and Row, 1963.

——. "Roosevelt, Norris, and the 'Seven Little TVA's.'" *Journal of Politics* 14 (1952): 418–41.

Levi, Edward H. *An Introduction to Legal Reasoning.* Chicago: University of Chicago Press, 1949.

Liebman, Ellen. *California Farmland: A History of Large Agricultural Landholdings.* Totowa, N.J.: Rowman and Allenheld, 1983.

Llewellyn, K. N. *The Bramble Bush: On Our Law and Its Study.* 1930. Reprint, New York: Oceana, 1951.

Louisiana Engineering Society. *Government Control With Cooperation of Riparian States and Cities in the Construction and Maintenance of Levees, Bank Revetments and Channel Enlargements for the Rivers of the Mississippi Valley as Adequate Means of Flood Protection Resulting in Increased Wealth and Stability of Investment. Addresses Before the Louisiana Engineering Society, June 25, 1912.* New Orleans: Louisiana Engineering Society, 1912.

Lowi, Theodore J. "American Business, Public Policy, Case-Studies, and Political Theory." *World Politics* 16 (1964): 677–715.

——. "Four Systems of Policy, Politics, and Choice." *Public Administration Review* 32, no. 4 (1972): 298–310.

——. "The Welfare State, the New Regulation, and the Rule of Law." In *Distributional Conflicts in Environmental-Resource Policy,* edited by Allan Schnaiberg, Nicholas Watts, and Klaus Zimmermann, 109–49. New York: St. Martin's Press, 1986.

Lowi, Theodore J., and Edward J. Harpham. "Political Theory and Public Policy: Marx, Weber, and a Republican Theory of the State." In *Contemporary Empirical Political Theory,* edited by Kristen Renwick Monroe, 249–78. Berkeley: University of California, 1997.

Maass, Arthur. *Muddy Waters: The Army Engineers and the Nation's Rivers.* Cambridge: Harvard University Press, 1951.

Mann, Michael. *The Sources of Social Power.* Vol. 1: *A History of Power From the Beginning to A.D. 1760.* Cambridge: Cambridge University Press, 1986.

——. *The Sources of Social Power.* Vol. 2: *The Rise of Classes and Nation-States, 1760–1914.* Cambridge: Cambridge University Press, 1993.

Mann, Ralph. *After the Gold Rush: Society in Grass Valley and Nevada City, California, 1849–1870.* Stanford: Stanford University Press, 1982.

Mann, Susan. *Agrarian Capitalism in Theory and Practice.* Chapel Hill: University of North Carolina Press, 1990.

Marcus, Melvin G. "Environmental Policies in the United States." In *Environmental Policies: An International Review,* edited by Chris C. Park, 45–75. London: Croom Helm, 1986.

May, Philip Ross. *Origins of Hydraulic Mining in California.* Oakland: Holmes Book Co., 1970.

McConnell, Grant. *The Decline of Agrarian Democracy.* New York: Atheneum, 1959.

——. *Private Power and American Democracy.* New York: Alfred A. Knopf, 1966.

McLeod, Doug. " 'Historic' Water Pact Rooted in Compromise." *Manteca (California) Bulletin*, 16 December 1994, A1, A4.

McPhee, John. "Annals of the Former World: Assembling California—1." *New Yorker*, September 1992, 36–68.

——. "Annals of the Former World: Assembling California—2." *New Yorker*, September 1992, 44–84.

——. "Atchafalaya." In *The Control of Nature*, 3–92. New York: Farrar Straus Giroux, 1986.

Merchants' Exchange of St. Louis Mississippi River Improvement Committee. *The Improvement of the Mississippi River. An Address Delivered at St. Louis, January 26, 1884, and Two Articles Originally Published in the North American Review by Robert S. Taylor of the Mississippi River Commission*. St. Louis: Merchants' Exchange of St. Louis [1884?].

Miliband, Ralph. *The State in Capitalist Society*. New York: Basic Books, 1969.

Mills, William Harrison. "The Hydrography of the Sacramento Valley." *California State Board of Trade Bulletin* 11 [1904?].

Minicucci, Stephen. "The 'Cement of Interest': Interest-Based Models of Nation-Building in the Early Republic." *Social Science History* 25 (2001): 247–74.

Mississippi River Convention Quincy. *Proceedings of the Mississippi River Convention Held in the Rooms of the Young Men's Business Association, Quincy, Ill [sic], Thursday, October 13th, 1887. Official Report*. Quincy, Ill.: The Convention, 1887.

Mississippi River Improvement Convention. *A Memorial to Congress to Secure an Adequate Appropriation For a Prompt and Thorough Improvement of the Mississippi River*. St. Louis: Mississippi River Improvement Convention, St. Paul, Executive Committee, 1877.

——. *Official Report of the Proceedings of the Mississippi River Improvement Convention. St. Louis, Missouri, October 26, 27, and 28, 1881*. St. Louis: Mississippi River Improvement Convention, 1881.

Mississippi River Improvement Convention Dubuque 1866. *Proceedings of the Mississippi River Improvement Convention Held at Dubuque, Iowa, February 14th and 15th, 1866*. Dubuque: Mississippi River Improvement Convention, 1866.

Mississippi River Improvement and Levee Convention. "Resolutions of the Mississippi River Improvement and Levee Convention, Held at Vicksburg, Mississippi, April 30 and May 1, 1890, and Resolution Adopted by the Southern Press Association." In U.S. Senate Committee on Commerce. *Hearings Before the Committee on Commerce, U.S. Senate, In Relation to the Improvement of the Mississippi River*, 146–51. Washington, D.C.: Committee on Commerce, 1890.

Mississippi River Levee Association. *Flood Protection on the Lower Mississippi River, A National Problem: Public Sentiment as Expressed by Mayors of Cities and Commercial Organizations Throughout the United States*. Memphis: Mississippi River Levee Association [1916?].

Mississippi River Levee Association. *Mississippi River Levee Association* [pamphlet]. Memphis: Mississippi River Levee Association, n.d.

——. *Prominent Bankers Favoring Legislation by Congress to Prevent Floods*. Memphis: Mississippi River Levee Association, 1913.

——. *Protection from Overflow and Reclamation of Thirty Thousand Square Miles of Alluvial Lands of the Lower Mississippi River, a National Work*. Memphis: Mississippi River Levee Association [1913?].

Mitchell, Broadus. *Depression Decade: From New Era through New Deal, 1929–1941*. New York: Rinehart, 1947.

Mitchell, Timothy. "The Limits of the State: Beyond Statist Approaches and Their Critics." *American Political Science Review* 85, no. 1 (1991): 77–96.

——. "Society, Economy, and the State Effect." In *State/Culture: State-Formation After the Cultural Turn*, edited by George Steinmetz, 76–97. Ithaca: Cornell University Press, 1999.

Mooney, Patrick H. "Labor Time and Capitalist Development in Agriculture: A Reconsideration of the Mann-Dickinson Thesis." *Sociologia Ruralis* 22 (1982): 279–92.

Moore, Barrington, Jr. *Injustice: The Social Bases of Obedience and Revolt*. White Plains, N.Y.: M. E. Sharpe, 1978.

——. *Social Origins of Dictatorship and Democracy: Lord and Peasant in the Making of the Modern World*. Boston: Beacon, 1966.

Moore, Jamie W., and Dorothy P. Moore. *The Army Corps of Engineers and the Evolution of Federal Flood Plain Management Policy*. Boulder: Institute of Behavioral Science, University of Colorado, 1989.

Moorhead, Dudley. "Sectionalism in the California Constitution of 1879." *Pacific Historical Review* 12 (1943): 287–93.

Morgan, Arthur E. *Dams and Other Disasters: A Century of the Army Corps of Engineers in Civil Works*. Boston: Porter Sargent, 1971.

Morison, Samuel Eliot. *The Oxford History of the American People*. 3 vols. New York: Oxford University Press, 1972.

Nasar, Sylvia. "Damage Estimates of Midwest Flood Climbing Sharply." *New York Times*, 25 July 1993, A1, A18.

National Drainage Congress. *Official Proceedings of the Eighth Annual Meeting of the National Drainage Congress at St. Louis, Missouri, November 11, 12 and 13, 1919*. St. Louis: National Drainage Congress, 1919.

——. *Official Proceedings of the Sixth Annual Meeting of the National Drainage Congress, Cairo, Illinois, January 19, 20 and 21, 1916*. Cairo: National Drainage Congress, 1916.

National Rivers and Harbors Congress. *Proceedings of the Convention, National Rivers and Harbors Congress*. Cincinnati: National Rivers and Harbors Congress. 1906–11.

——. *Proceedings of the Convention, National Rivers and Harbors Congress.* Washington, D.C.: National Rivers and Harbors Congress. 1912–28.

——. *Bulletin.* Washington, D.C., National Rivers and Harbors Congress. 1920–31.

——. *Resolutions Adopted by the Annual Convention of the National Rivers and Harbors Congress.* Washington, D.C., National Rivers and Harbors Congress. 1935–36.

Nettels, Curtis. "The Mississippi Valley and the Constitution." *Mississippi Valley Historical Review* 11, no. 3 (1924): 332–57.

New York Board of Trade and Transportation. *Mississippi Valley Floods Relief: A National Obligation.* New York: New York Board of Trade and Transportation, 1914.

Northwestern Waterways Convention, St. Paul. *Official Proceedings of the Northwestern Waterways Convention Held at the City of St. Paul, September 3 and 4, 1885.* St. Paul: Northwestern Waterways Convention, 1885.

O'Neill, Karen M. "Why the TVA Remains Unique: Interest Groups and the Defeat of New Deal River Planning." *Rural Sociology* 67, no. 2 (2002): 163–82.

Orloff, Ann Shola. "The Political Origins of America's Belated Welfare State." In *The Politics of Social Policy in the United States,* edited by Margaret Weir, Ann Shola Orloff, and Theda Skocpol, 37–80. Princeton: Princeton University, 1988.

Park, W. L. *Conservation, Reclamation, Navigation.* Little Rock, Ark.: Lakes-to-the-Gulf Deep Waterway Association, 1912.

Parker, Walter. "Curbing the Mississippi." *The Nation,* 11 May 1927, 521–22.

Paul, Rodman Wilson. "The Great California Grain War: The Grangers Challenge the Wheat King." *Pacific Historical Review* 27 (1958): 331–49.

Pearcy, Matthew T. "After the Flood: A History of the 1928 Flood Control Act." *Journal of the Illinois State Historical Society* 95, no. 2 (2002): 172–201.

Perkins, Broderick. "Flood Insurance: What You Can Do to Cut Cost, Need." *San Jose Mercury News,* 18 October 1997, F1.

Peterson, Richard H. "The Failure to Reclaim: California State Swamp Land Policy and the Sacramento Valley, 1850–1866." *Southern California Quarterly* 56 (1974): 45–60.

Petulla, Joseph. *American Environmental History: The Exploitation and Conservation of Natural Resources.* San Francisco: Boyd and Fraser, 1977.

Pfeffer, Max J. "Social Origins of Three Systems of Farm Production in the United States." *Rural Sociology* 48, no. 4 (1983): 540–62.

Pisani, Donald J. *From the Family Farm to Agribusiness: The Irrigation Crusade in California and the West, 1850–1931.* Berkeley: University of California Press, 1984.

Polanyi, Karl. *The Great Transformation.* Boston: Beacon, 1944.

Rapids Convention [of 1851] Burlington. *Proceedings of the Rapids Convention,*

Held at Burlington, Iowa, on the 23rd and 24th of October, 1851. Burlington: Rapids Convention, 1852.

Reagan, Michael D., and John G. Sanzone. *The New Federalism.* 2nd ed. New York: Oxford University Press, 1981.

Reclamation and Drainage Association. [Report]. Sacramento, [1902?].

Reichley, A. James. *The Life of the Parties.* New York: Free Press, 1992.

Reisner, Marc. *Cadillac Desert: The American West and Its Disappearing Water.* New York: Viking, 1986.

Reuss, Martin. "Andrew A. Humphreys and the Development of Hydraulic Engineering: Politics and Technology in the Army Corps of Engineers, 1850–1950." In *The Engineer in America: A Historical Anthology from Technology and Culture*, edited by Terry S. Reynolds, 89–121. Chicago: University of Chicago Press, 1991.

——. *Designing the Bayous: The Control of Water in the Atchafalaya Basin, 1800–1995.* Alexandria, Va.: U.S. Army Corps of Engineers, 1998.

——. "The Development of American Water Resources: Planners, Politicians, and Constitutional Interpretation." In *Managing Water Resources Past and Present*, edited by Julie Trottier and Paul Slack, 51–72. Oxford: Oxford University Press, 2004.

Revkin, Andrew C. "Stockpiling Water for a River of Grass." *New York Times*, 26 March 2002, F1, F4.

Riche, C. S. *The Further Improvement of Our Inland Waterways. A Paper Read Before the Contemporary Club, Davenport, Iowa, March 19, 1908.* Davenport, Iowa: Contemporary Club, 1908.

Rightor, Henry. *Standard History of New Orleans.* Chicago: Lewis Publishing, 1890.

River and Canal Improvement Convention Davenport. *Proceedings of the River and Canal Improvement Convention, Davenport, Iowa, 1881.* River and Canal Improvement Convention, Davenport, Iowa, 1881.

River and Harbor Improvement Convention of 1885 Tuscaloosa. *Memorial and Proceedings of the River and Harbor Improvement Convention, Tuscaloosa, Alabama, November 17, 1885.* Cincinnati: River and Harbor Improvement Convention, Tuscaloosa, 1886.

River Improvement and Drainage Association of California. *Report of the Committee of Twenty-four of the River Improvement and Drainage Association of California.* San Francisco: River Improvement and Drainage Association, 1902.

——. *Bulletin.* San Francisco: River Improvement and Drainage Association of California, 1904–8.

River Improvement Convention Quincy. *Memorial of the Quincy River Improvement Convention, October 15, 1879 to the Congress of the United States, in Regard to the Improvement of the Mississippi River and its Tributaries.* St. Louis: River Improvement Convention, 1879.

River Improvement Convention St. Louis. *Proceedings of the River Improvement Convention, Held In St. Louis, February 12 and 13, 1867*. St. Louis: Union Merchants' Exchange of St. Louis, 1867.

River Improvement Meeting Chicago 1896. *Chicago! Where Railroad Traffic and Lake Transportation Meet. Snap Shots From Proceedings of the River Improvement Meeting, at Great Northern Hotel, March 13, 1896*. Chicago: Association for the Improvement of the Chicago River, 1896.

Robbins, Roy M. *Our Landed Heritage*. Lincoln: University of Nebraska Press, 1976.

Robinson, Michael C. *Water for the West: The Bureau of Reclamation, 1902–1977*. Chicago: Public Works Historical Society, 1979.

Robinson, William W. *Land in California: The Story of Mission Lands, Ranchos, Squatters, Mining Claims, Railroad Grants, Land Scrip [and] Homesteads*. Berkeley: University of California Press, 1948.

Rodgers, Daniel T. "In Search of Progressivism." *Reviews in American History* December (1982): 113–31.

Rogers, Peter. *America's Water: Federal Roles and Responsibilities*. Cambridge, Mass.: Twentieth Century Fund, MIT Press, 1993.

Rohe, Rand E. "Man as Geomorphic Agent: Hydraulic Mining in the American West." *Pacific Historian* 27, no. 1 (1983): 5–16.

Rohe, Randall. "Man and the Land: Mining's Impact in the Far West." *Arizona and the West* 28 (1986): 299–338.

Rohrbough, Malcolm J. *The Land Office Business: The Settlement and Administration of American Public Lands, 1789–1837*. Belmont: Wadsworth, 1990.

Rokkan, Stein. *Cities, Elections, Parties*. Oslo: Universitetsforlaget, 1970.

Roy, William G. *Socializing Capital: The Rise of the Large Industrial Corporation in America*. Princeton: Princeton University Press, 1997.

Rubinson, Richard. "Political Transformation in Germany and the United States." In *Social Change in the Capitalist World Economy*, edited by Barbara Hockey Kaplan, 39–73. Beverly Hills: Sage, 1978.

Sacramento River Drainage District. *Report of the Board of Commissioners of the Sacramento River Drainage District to the Governor of California*. Sacramento: State Office, 1879.

Sacramento Valley Anti-Debris Association. *A Brief Outline of the Economic Features of Hydraulic Mining. Prepared for the 1927 State Legislature 1928, Water Conservation Committee*. Yuba City, California: Sacramento Valley Anti-Debris Association [1927?].

Sacramento Valley Development Association. *Five Years' Review: Being a Report of the Executive Committee of the Sacramento Valley Development Association*. Sacramento, 1905.

Sahlins, Peter. *Boundaries: The Making of France and Spain in the Pyrenees*. Berkeley: University of California Press, 1989.

San Joaquin and Sacramento River Improvement Association. *Articles of Association and By-laws.* San Francisco [1909?].

Savannah Valley Convention. *Memorial to the Congress of the United States, February 1, 1888.* Augusta, Ga.: Savannah Valley Convention, 1888.

Saxon, Lyle. *Father Mississippi.* New York: The Century Co., 1927.

Schaeffer, Robert K. *Severed States: Dilemmas of Democracy in a Divided World.* Lanham, Md.: Rowman and Littlefield, 1999.

Schaffer, Daniel. "Managing the Tennessee River: Principles, Practice, and Change." *Public Historian* 12 (1990): 7–29.

Schapsmeier, Edward L., and Frederick H. Schapsmeier. *Political Parties and Civic Action Groups.* Westport, Conn.: Greenwood Press, 1981.

Scheiber, Harry N. "Federalism and the American Economic Order, 1789–1910." *Law and Society Review* 10 (1975): 57–118.

Schreurs, Miranda A. "Conservation, Development, and State Sovereignty: Japan and the Tropical Forests of Southeast Asia." In *State Sovereignty: Change and Persistence in International Relations,* edited by Sohail H. Hashmi, 181–204. University Park: Pennsylvania State University Press, 1997.

Seip, Terry L. *The South Returns to Congress: Men, Economic Measures, and Intersectional Relationships, 1868–1879.* Baton Rouge: Louisiana State University Press, 1983.

Selznick, Philip. *TVA and the Grass Roots: A Study in Politics and Organization.* 1949. Reprint, Berkeley: University of California Press, 1980.

Shallat, Todd. "Engineering Policy: The U.S. Army Corps of Engineers and the Historical Foundation of Power." *Public Historian* 11, no. 3 (1989): 7–27.

——. *Structures in the Stream: Water, Science, and the Rise of the U.S. Army Corps of Engineers.* Austin: University of Texas Press, 1994.

Shanks, Bernard. *This Land Is Your Land: The Struggle to Save America's Public Lands.* San Francisco: Sierra Club, 1984.

Shannon, Fred A. *The Farmer's Last Frontier: Agriculture, 1860–1897.* 1945. Reprint, New York: Holt, Rinehart and Winston, 1966.

Shinn, Charles Howard. *Mining Camps: A Study in Frontier Government.* 1885. Reprint, New York: Alfred A. Knopf, 1948.

Short, Lloyd M. *The Development of National Administrative Organization in the United States.* Baltimore: Johns Hopkins University Press, 1923.

Sillers, Walter, Sr. "Flood Control in Bolivar County, 1838–1924." *Journal of Mississippi History* 9, no. 1 (1947): 3–20.

Sitterson, J. Carlyle. *Sugar Country: The Cane Sugar Industry in the South, 1753–1950.* Lexington: University of Kentucky Press, 1953.

Skocpol, Theda. *Protecting Soldiers and Mothers: The Political Origins of Social Policy in the United States.* Cambridge: Belknap Press of Harvard University Press, 1992.

——. *States and Social Revolutions: A Comparative Analysis of France, Russia, and China.* Cambridge: Cambridge University Press, 1979.

Skocpol, Theda, and Kenneth Finegold. "State Capacity and Economic Intervention in the Early New Deal." *Political Science Quarterly* 97, no. 2 (1982): 255–78.

Skocpol, Theda, and John Ikenberry. "The Political Formation of the American Welfare State in Historical and Comparative Perspective." *Comparative Social Research* 6 (1983): 87–148.

Skowronek, Stephen. *Building a New American State: The Expansion of National Administrative Capacities, 1877–1920*. Cambridge: Cambridge University Press, 1982.

Smith, Kathie A. "Deal OK'd for First River Plan Since 1911." *Modesto [California] Bee*, 3 March 1998, B1–B2.

St. Louis Chamber of Commerce. *Proceedings of the St. Louis Chamber of Commerce, in Relation to the Improvement of the Navigation of the Mississippi River and Its Principal Tributaries and the St. Louis Harbor. With a Statement Submitted by A. B. Chambers to the Chamber*. St. Louis: St. Louis Chamber of Commerce, 1842.

St. Louis Citizens. *Memorial of the Citizens of St. Louis, Missouri, to the Congress of the United States: Praying an Appropriation for Removing the Obstruction to the Navigation of the Western Rivers, for the Improvement of the St. Louis Harbor and for Other Purposes*. St. Louis: Chambers and Knapp, 1844.

Starling, William. *The Floods of the Mississippi River: Including an Account of their Principal Causes and Effects: And a Description of the Levee System and Other Means Proposed and Tried for the Control of the River, With a Particular Account of the Great Flood of 1897*. New York: Engineering News Publishing Co., 1897.

State of California. Swamp Land Commissioners. *First Annual Report of the Swamp Land Commissioners*. Sacramento: Swamp Land Commissioners, 1862.

Stein, Robert M., and Kenneth Bickers. *Perpetuating the Pork Barrel: Policy Subsystems and American Democracy*. New York: Cambridge University Press, 1995.

Steinmetz, George. "Introduction: Culture and the State." In *State/Culture: State-Formation after the Cultural Turn*, edited by George Steinmetz, 1–49. Ithaca: Cornell University Press, 1999.

Stine, Jeffrey K. "United States Army Corps of Engineers (CE)." In *Government Agencies: The Greenwood Encyclopedia of American Institution*, edited by Donald R. Whitnah, 513–16. Westport, Conn.: Greenwood Press, 1983.

Stokes, Anson Phelps, Jr. "The Congressional Pork Barrel." *Harper's Weekly*, 22 March 1913.

Stone, Lawrence. "Social Mobility in England, 1500–1700." *Past and Present* 33 (1966): 16–55.

Swain, Donald. *Federal Conservation Policy, 1921–1933*. Berkeley: University of California Press, 1963.

Taylor, Robert S. "The Subjugation of the Mississippi." *North American Review* 136 (1883): 212–22.

Thompson, John. *The Settlement Geography of the Sacramento-San Joaquin Delta, California*. PH.D. dissertation, Stanford University, 1957.

Thompson, John, and Edward A. Dutra. *The Tule Breakers: The Story of the California Dredge*. Stockton, Calif.: Stockton Corral of Westerners and University of the Pacific, 1983.

Tilly, Charles. *As Sociology Meets History*. Orlando: Academic Press, 1981.

——. *Coercion, Capital, and European States*, AD 990–1990. Cambridge, Mass.: Basil Blackwell, 1990.

——. *The Contentious French*. Cambridge: Belknap Press of Harvard University Press, 1986.

——. "Epilogue: Now Where?" In *State/Culture: State-Formation After the Cultural Turn*, edited by George Steinmetz, 407–19. Ithaca: Cornell University Press, 1999.

——. "War Making and State Making as Organized Crime." In *Bringing the State Back In*, edited by Peter B. Evans, Dietrich Rueschemeyer, and Theda Skocpol, 169–91. Cambridge: Cambridge University Press, 1985.

Tocqueville, Alexis de. *The Old Regime and the French Revolution*. Translated by S. Gilbert. Garden City, N.Y.: Doubleday, 1955.

Tompkins, Frank H. *Riparian Lands of the Mississippi River*. New Orleans: Mississippi River Improvement Levee Association, 1901.

Townsend, McD. *Flood Control of the Mississippi River. Address Before the National Drainage Congress. St. Louis, Missouri, April 11, 1913*. Memphis: Mississippi River Levee Association, 1913.

Trotsky, Leon. *The Russian Revolution*. 1932. Reprint, Garden City, N.Y.: Doubleday Anchor, 1959.

Twain, Mark [Samuel Clemens]. *Life on the Mississippi*. 1883. Reprint, New York: Modern Library, 1994.

U.S. Army Corps of Engineers Historical Division and Public Affairs Office. *The History of the* U.S. Army Corps of Engineers. Fort Belvoir, Va.: U.S. Army Corps of Engineers, 1986.

U.S. Army Corps of Engineers, Lower Mississippi Valley Division. *Water Resources Development in Louisiana, 1989*. New Orleans: U.S. Army Corps of Engineers, Lower Mississippi Valley Division, 1989.

U.S. Army Corps of Engineers, CDC California Debris Commission. *Enacts the 1902 Rivers and Harbors Act Plan for Yuba River*. House document 431, 56th Congress, 1st session, 1902.

U.S. Army Corps of Engineers, Chief of Engineers. *Annual Report 1896 Upon the Improvement of San Joaquin and Sacramento Rivers and Tributaries, and of Rivers and Harbors in California North of San Francisco. W. H. Heuer, Major, Corps of Engineers and Cassius E. Gillette, Captain. (Appendix SS of Annual*

Report). Washington, D.C.: U.S. Army Corps of Engineers, Chief of Engineers, 1896.

——. *Annual Report*. Washington, D.C.: U.S. Army Corps of Engineers, Chief of Engineers, 1918.

U.S. Army Corps of Engineers, Sacramento District. Water Resources Planning Branch Investigations Section A. *Sacramento River Aerial Atlas*. Sacramento: U.S. Army Corps of Engineers, Sacramento District, in cooperation with the California Department of Water Resources, 1980.

U.S. Army Corps of Engineers, South Pacific Division. *Water Resources Development, California, 1979*. San Francisco: Corps South Pacific Division, 1979.

U.S. Army, Office of the Chief of Engineers. *Control of Floods in the Alluvial Valley of the Lower Mississippi River*. House document 798. 71st Cong., 3rd sess., 1931.

U.S. Congress. *Congressional Record*. 64th Cong., 1st sess., 10 May 1916.

U.S. Department of the Interior. General Land Office. *Annual Reports of the Department of the Interior. Report of the Commissioner of the General Land Office*. House doc. 90 (64-1) 6991, vol. 37, 1915. Washington, D.C.: Department of Interior.

U.S. Engineer Department. *Estimates of Cost of Examinations, etc. of Streams Where Power Development Appears Feasible*. House document 308. 69th Cong., 1st sess., 1926.

U.S. House of Representatives, Committee on Flood Control. *Hearings on Mississippi River Floods*. 64th Cong., 1st sess, 1916.

U.S. House of Representatives, Committee on the Public Lands. *Swamp and Overflowed Lands*. House report no. 347, 50th Cong., 1st sess. (to accompany HR 6897), 1888.

U.S. Senate Committee on Commerce. *Flood Control Act of 1936. An Act Authorizing the Construction of Certain Public Works on Rivers and Harbors for Flood Control and Other Purposes*. Confidential hearings. HR 8455. 74th Cong., 2nd sess. March 19 and 25, 1936.

U.S. War Department, Board of Engineers [Biggs Commission]. *Mining Debris, California*. House of Representatives doc. no. 261. 51st Cong., 2nd sess. February 20, 1891.

U.S. War Department, Commission of Engineers. *Report of the Commission of Engineers Appointed to Investigate and Report a Permanent Plan for the Reclamation of the Alluvial Basin of the Mississippi River Subject to Inundation*. House of Representatives executive doc. no. 127. 43rd Cong., 2nd sess., 1875.

Union Merchants' Exchange of St. Louis. *Memorial of the Union Merchants' Exchange of St. Louis to the Forty-third Congress of the United States. A Statement of the Necessities of the People of the Mississippi Valley, in Respect to the Transportation of Their Products, and the Improvement of Their Natural Channels of Commerce*. St. Louis: Union Merchants' Exchange of St. Louis, 1874.

Union Merchants' Exchange of St. Louis. *Report of the Committee on Improvement of the Mississippi River and Tributaries. St. Louis, 15th December, 1865*. St. Louis: Union Merchants' Exchange of St. Louis [1865?].

Walton, John. *Western Times and Water Wars: State, Culture, and Rebellion in California*. Berkeley: University of California Press, 1992.

Water for People, Water for Life: The United Nations World Water Development Report, Executive Summary. Paris: UNESCO, 2003.

Weaver, John C. "Beyond the Fatal Shore: Pastoral Squatting and the Occupation of Australia, 1826 to 1852." *American Historical Review* 101 (1996): 981–1007.

Weber, Max. *Economy and Society*. 1956. Reprint, Berkeley: University of California Press, 1978.

——. "Politics as a Vocation." In *From Max Weber: Essays in Sociology*. 1919. Reprint edited by H. H. Gerth and C. Wright Mills, 77–128. New York: Oxford University Press, 1946.

Weigley, Russell F. *History of the United States Army*. Enlarged edition. Bloomington: Indiana University Press, 1984.

Wells, Harry L. *History of Nevada County, California*. Oakland: Thompson and West, 1880.

Wells, J. Maddison. *The National Importance of Rebuilding and Repairing the Levees, on the Banks of the Mississippi . . .* N.p., n.d. [c.1860s].

Welsh, Michael M. "Beyond Designed Capture: A Reanalysis of the Beginnings of Public Range Management, 1928–38." *Social Science History* 26 (2002): 347–91.

White, Gilbert F. "Human Adjustment to Floods: A Geographical Approach to the Flood Problem in the United States." University of Chicago, Department of Geography, Research paper no. 29, 1942. Chicago, 1945.

White, Gilbert F., and John Eric Edinger. *Choice of Adjustment to Floods*. Department of Geography research paper no. 93. Chicago: Department of Geography, University of Chicago, 1964.

Wiebe, Robert H. *The Search for Order, 1877–1920*. New York: Hill and Wang, 1967.

Wilentz, Sean. "Society, Politics, and the Market Revolution, 1815–1848." In *The New American History*, edited by Eric Foner, 51–71. Philadelphia: Temple University Press, 1990.

Williams, Albert N. *The Water and the Power: Development of the Five Great Rivers of the West*. New York: Duell, Sloan and Pearce, 1951.

Williams, Mentor L. "The Background of the Chicago River and Harbor Convention, 1847." *Mid-America* 30 (1948): 219–32.

——. "The Chicago River and Harbor Convention, 1847." *Mississippi Valley Historical Review* 35 (1949): 607–26.

Wilson, Thomas M., and Hastings Donnan, eds. *Border Identities: Nation and*

State at International Frontiers. Cambridge: Cambridge University Press, 1998.

Winston, James E. "Notes on the Economic History of New Orleans, 1803–1836." *Mississippi Valley Historical Review* 11, no. 2 (1924): 200–26.

Wolf, Eric R. *Europe and the People Without History*. Berkeley: University of California Press, 1982.

Woodward, C. Vann. *Origins of the New South, 1877–1913*. Baton Rouge: Louisiana State University Press, 1951.

——. *Reunion and Reaction: The Compromise of 1877 and the End of Reconstruction*. Boston: Little Brown, 1966.

Woolfolk, George Ruble. *The Cotton Regency: The Northern Merchants and Reconstruction, 1865–1880*. New York: Bookman Associates, 1958.

Worster, Donald. *Rivers of Empire: Water, Aridity, and the Growth of the American West*. New York: Pantheon, 1985.

Wyatt, Dennis. "This Year's Flooding Is 1 in 100." *Manteca (California) Bulletin*, 17 January 1997, A4.

Young, James T. "The Origins of New Deal Agricultural Policy: Interest Groups' Role in Policy Formation." *Policy Studies Journal* 21, no. 1 (1993): 190–209.

Index

KAREN M. O'NEILL

is an assistant professor of human ecology

at Rutgers University.

Library of Congress Cataloging-in-Publication Data

O'Neill, Karen M., 1959–

Rivers by design : state power and the origins of U.S.

flood control / Karen M. O'Neill.

p. cm.

Includes bibliographical references and index.

ISBN 0-8223-3760-6 (cloth : alk. paper)—

ISBN 0-8223-3773-8 (pbk. : alk. paper)

1. Flood control—Government policy—United States.

2. River engineering—Government policy—United

States. 3. River engineering—Mississippi River.

4. Flood control—California—Sacramento River.

5. River engineering—California—Sacramento River.

6. Flood control—Mississippi River. 7. State rights.

I. Title.

TC530.055 2006

363.34'9360973—dc22 2005033598